THE TELEPHONE EXCHANGE AT COPENHAGEN, DENMARK.

This Exchange is equipped with Ericsson multiple switchboards, over 8,000 subscribers' lines being connected. The capacity of the board is 10,000 subscribers, with distributing facilities for 12,000. The hall shown in the plate is the largest single switchboard room in the world.

A B C

of the

TELEPHONE

A Practical and Useful Treatise for
Students and Workers in Telephony,
Giving a Review of the Development of
the Industry to the Present Date, and Full
Descriptions of Numerous Valuable
Inventions and Appliances, Together
with Very Many Illustrations, Diagrams
and Tables ✍ ✍ ✍ ✍ ✍ ✍ ✍ ✍ ✍

By

JAMES E. HOMANS, A.M.

THEO. AUDEL & CO., Publishers
63 Fifth Ave., Cor. 13th St., N. Y.
1901

ALEXANDER GRAHAM BELL,
Inventor of the Magnet Telephone.

PREFACE.

As is indicated by the title, this book is an elementary and general treatise. It deals with the telephone industry and the leading features of its development to the present time, in such a fashion and in such language as to enable the student and the non-professional reader to derive an intelligent idea of broad facts and general principles. Of course, with such an aim in view, it cannot be claimed that the book will furnish all the details and formulæ that may be required by the thoroughly equipped telephonist. Such facts, at best, can be mastered only after practical experience and careful training ; and, without this, no book can guarantee a thorough education in the profession of constructing and maintaining telephone plants.

It is claimed, however, that the information given is sufficiently full and precise to render the work useful as reference for the student or worker in telephony, and to furnish a good foundation for more extended study of the subject. Many excellent treatises on technical topics, in the effort to set forth all there is to be known, surround the simplest facts with such a maze of formulæ and difficult details that the beginner can only with the utmost difficulty derive any kind of an intelligent idea of the facts he must master. Such a method is opposed to the most enlightened theories of education at the present day, which prescribe a plain and intelligible review of the subject before an attempt is made to present a more advanced and technical discussion. This is the function the A B C of the Telephone is designed to serve.

In the preparation of the work several noted authorities have been consulted, and their assistance is frequently acknowledged. In several cases the treatment of subjects closely parallels that used by other writers, with the exception that the plain facts of the case are given in language as far as possible adapted to non-professional comprehension, omitting such details as would be useful only to the most advanced students of the subject.

Numerous manufacturers of telephones and supplies have rendered valuable assistance in enabling full and authoritative descriptions of their products and in furnishing cuts and illustrations suitable to accompany them. As it happens to be the case in telephony, as also in about every other mechanical and electrical industry, that inventors and manufacturers are closely associated, the most important inventions to be described are proprietary articles and must be duly credited. In this connection acknowledgments are due to the Western Telephone Construction Co., the Sterling Electric Co., the Keystone Electric Telephone Co., the Holtzer-Cabot Electric Co., the Couch & Seeley Co., the Ericsson Telephone Co., the Century Telephone Construction Co., and the Connecticut Telephone and Electric Co.

Mr. Charles E. Monroe, of the Ericsson Telephone Co., has rendered important assistance in the preparation of this book by several suggestions of value and by reading the proofs. His kindness is hereby acknowledged.

CONTENTS.

CHAPTER FIFTEEN.

CHAPTER SIXTEEN.

CHAPTER SEVENTEEN.

CHAPTER EIGHTEEN.

PRIVATE TELEPHONE LINES AND INTERCOMMUNICATING
SYSTEMS ; COMMON RETURN CIRCUITS.

CHAPTER NINETEEN.

PRIVATE TELEPHONE LINES AND INTERCOMMUNICATING SYS-
TEMS ; FULL METALLIC CIRCUITS.

CHAPTER TWENTY.

LARGE PRIVATE SYSTEMS AND AUTOMATIC EXCHANGES

CHAPTER TWENTY-ONE.

DEVICES FOR PROTECTING TELEPHONE APPARATUS FROM ELEC-
TRICAL DISTURBANCES.

CHAPTER TWENTY-TWO.

CHAPTER TWENTY-THREE.

CHAPTER TWENTY-FOUR.

CHAPTER TWENTY-FIVE.

CHAPTER TWENTY-SIX.

PA

A B C

OF

THE TELEPHONE.

CHAPTER ONE.

THE TELEPHONE APPARATUS AND ITS OPERATION.

The Telephone.—The telephone is an instrument for the transmission of articulate speech by the electric current. So it was described in the application and specifications on which, in 1876, Alexander Graham Bell was granted letters patent for his magnet telephone; and because of this fortunate form of words, which covered the process as well as the device, he was able to maintain a complete monopoly of the telephone business, until the expiration of his patent rights, seventeen years later.

Inventors of the Telephone.—Previous to the date of Bell's patent several experimenters had hit upon the idea of transmitting sound by the electric current—among these was Elisha Grey—but as must seem strange at the present day all these persons regarded their instruments as little more than scientific toys, or curiosities for exhibition purposes. Such, indeed, has been the experience of most of our great modern inventions; at first people doubted their possibility, then their utility, and were finally convinced of both only by practical demonstration. Now that the telephone has fully established its claim to attention, and become a useful, even indispensable, article in business and other concerns of life, it is desirable that everyone should have some knowledge of its construction, operation and the theory upon which it depends. To serve such a purpose is the object of this book.

Derivation of the Word.—The word, telephone, is formed from the two Greek words, *tele*, afar, and *phonein*, to sound, or to speak; and hence means the "far-speaker." Just so, we have the word, telegraph, from *tele* and *graphein*, to write, with the meaning, "far-writer," and telescope, from *tele* and *skopein*,

to see, meaning "far-seer." The word was in use many years before Bell's invention; in fact, as early as 1820, when Charles Wheatstone, one of the first experimenters on the theory of the electric telegraph, used it as the name of his device for transmitting the sound of the voice to a considerable distance along a wooden rod. This contrivance proved more efficient than the "lovers' telegraph," as it was called, the simplest form of acoustic telephone, consisting of two diaphragms closing, each an end of a tin can, connected by a taut string, which conveys the sound spoken into either can to the opposite end of the line. The "lovers' telegraph" was described in a book written by Robert Hooke, an English scientist, as early as 1667.

FIG. 1.—Usual form of Telephone Apparatus.

The Apparatus.—The most familiar form of modern telephone apparatus is given in Fig. 1, which shows its various parts to advantage. These are: a mounting-board bearing a closed box at top and bottom, a call-bell apparatus of two gongs to be operated by a crank shown at the right of the upper box; the *receiving instrument* hanging on a forked hook at the left; the *transmitting instrument* mounted below on a "rocker arm" having an up and down movement.

How to Use the Telephone.—To use the apparatus it 'cessary, first, to briskly turn the crank at the right of the upper box. This causes the bells to ring, and also sends a current along the line, which calls the operator at the central office. After "ringing up," remove the *receiving instrument* from its hook and apply it to the ear; the hook relieved of the weight of the receiver, immediately springs up. Having applied this instrument to the ear, you will hear the "hello girl" at "central," asking, "what number?" The names of all persons or firms having telephones are printed in the "Telephone Directory," which is to be found at every station, and each name has a number, by which the station may be called. Having previously found the number of the person with whom you wish to converse, you give it by *speaking into the transmitting instrument, never into the receiver.* As soon as connections have been made between your line and his, in a manner to be explained later, you begin the conversation, always speaking into the *transmitter* and hearing the answers from the other end through the *receiver.* On completing your conversation, you again hang the receiver on the hook, which is pulled downward by the weight, and having done this, "ring off" by a few rapid turns of the handle, again ringing the bells and thus informing "central" that you are through with the line.

FIG. 2.—The Generator Box showing Crank.

This is a simple act, and one which some of us perform many hundred times in a year; yet very few ever stop to consider how delicate and complicated is the machinery used, or by what wonderful, almost miraculous, processes the sound of the voice is thus transmitted over a line of wire, maybe a thousand miles long.

The Generator Box.—To gain some knowledge of the various parts of a working telephone, in order that the reader may know of what we speak when each is described in detail, we will dissect an apparatus showing views of each separate contrivance, and giving names and uses. Thus, Fig. 3 shows the interior of the box we have already seen attached to the upper end of the mounting-board in Fig. 1. The door, as shown, opens on hinges at its right-hand edge, thus concealing the crank of the call-bell previously mentioned. On the door, just back of the bell gongs, is to be seen a small frame support-

FIG. 3.—Interior of the generator box, showing the magneto-generator, switch-hook, and the magnets of the call bell.

ing an object resembling an ordinary spool, or rather two of them. This is an electro-magnet, and its function is to ring the bell when an electric current is passed through it, on the same principle as is seen in the electric door bell.

The Generator.—The object within the box, which resembles three upright posts, is the magneto-generator, consisting of three "horse-shoe" magnets, here shown in side view. Through their legs passes an "armature," a kind of metal reel wound about with a long coil of wire. When this is turned by means of the crank, previously mentioned, an electric current is produced, or "generated," as the term is. This machine is, in fact, a small dynamo.

The Calling Current.—The current passes from the generator through a wire attached to one of the hinges; thence to the bell-magnet; thence back again through a wire attached to the other hinge, and out upon the line wire to the central station, where it moves a signal to inform the operator that your station is calling.

FIG. 4.—One form of binding post. The screw at its end is intended for contact with the electrical conductor, and the line wire is inserted in the hole through the centre, where it is held tight by the thumb screw.

The line wires are attached to two of the binding posts shown at the top of the box in Fig. 2.

FIG. 5.—**Single Pole** Receiver. Shell made of hard rubber. Wires attached to the binding posts at the tail end.

The Hook Switch.—Directly in front of the magneto-generator is a lever, which passes through the side of the box and ends in the forked hook supporting the receiving instrument, as shown in Fig. 1. It is known as the "hook-switch," and its use, as will be explained in a later chapter, is to switch the talking instruments into circuit, when it is in the raised position, and to break that circuit and switch in the call-bell generator again, when the receiver is restored to its hook.

The Receiver.—Fig. 5 represents the receiver. At one end we see two binding-posts to receive wires which connect with the two binding-posts at the left of the row of six that are placed directly beneath the generator-box. The receiver is a magnet telephone of the kind invented and patented by Bell. Although many variations of the original instrument have since been devised by experimenters and manufacturers, none have improved materially on the original. The other speaking instrument, the transmitter, is made on a different principle,

FIG. 6.—The Transmitter mounted on Rocker Arm, attached to the Induction Coil Box at its base.

being a carbon microphone (or " sound-magnifier ") of the kind first devised by Mr. Edison. Both instruments, together with the many variations and improvements, will be fully explained in the proper place. We are at present concerned with their outer appearance, and the functions they serve in the practical operation of the apparatus.

The Transmitter and Induction Coil.—The usual form of the transmitter, enclosed in a semi-circular, or a cylindrical, case, surmounted by a bell-shaped mouthpiece, and supported at the end of a movable arm, is shown in Fig. 6. Wires run from it to the iron box at the base of the arm, and from this again other wires run from the four binding-posts shown in the figure, through the back of the mounting-board, to the remaining four binding-posts of the six beneath the generator-box.

The iron box contains an induction coil, a very useful and wonderful piece of apparatus, whose principles would require a book for discussion. Fig. 7 shows the form usual to telephones. The connections of the four wires just mentioned may

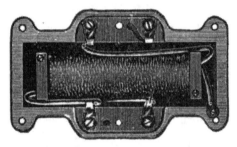

FIG. 7.—Induction coil mounted in the base of transmitter arm, showing wiring connections.

be seen in Fig. 8, two at either end. Two of them are connected, through the transmitting instrument, with a voltaic, or chemical, battery contained in the box at the base of the mounting-board, as shown in Fig. 1. The other two are directly

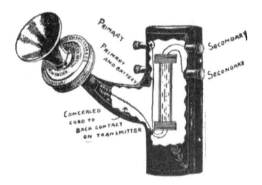

FIG. 8.—Section of transmitter arm and coil box, showing connections between the transmitter and induction coil.

connected to the line wires and carry the current used in transmitting the sounds spoken into the mouthpiece of the instrument.

This brief description will give a good general idea of

various parts of an ordinary telephone apparatus.   To fully
understand the use of each one, and why all are "assembled"
into one piece of machinery will involve a careful study of the
whole theory of the telephone.   This study, as we shall see,
will require us to cover a wide range of scientific facts.   For it
is true that a complete understanding of telephony demands a
good working knowledge of nearly every other branch of
electrical industry.

FIG. 9.—Form of Telephone Apparatus, manufactured in Sweden.   The magneto-
generator and induction coil are enclosed in the box below the call bells; the battery
cells, in the cupboard at the base of the backboard.   The transmitter is mounted at the
top.

CHAPTER TWO.

A BRIEF SURVEY OF THE THEORY OF SOUND, NECESSARY TO AN UNDERSTANDING OF THE TELEPHONE.

The Transmutation of Energy.—Among the many marvelous discoveries of modern science is what is known as the "transmutation" of energy or force, that is, the fact that one force may under the proper conditions be changed into any other. "Transmutation" means "changing about." Thus heat may be transformed into electrical activity and chemical energies may become either the one or the other. Heat may result from friction (rubbing), or from chemical action, as we see in fire, where the process by which a substance is reduced to ashes is called "combustion." Again heat, and light also, result from electricity. In short, scientists tell us that heat, light and sound are all one and the same thing, differing only in degree. Thus, when we feel that a body is hot, warm or cold; or when we see that light is brilliant, bright or dim; or when we hear a sound, and know that it is loud or indistinct, or differs from other sounds in representing a different musical note, we are only perceiving one fact, under different forms, and by different senses.

The Solar Spectrum.—A good illustration of this fact is to be found by comparing the octave of a musical instrument with the solar spectrum, as seen in a rainbow, or when sunlight shines through a prism. Very different things indeed these may seem to most observers, but science compares them. The solar spectrum shows seven colors—violet, indigo, blue, green, yellow, orange and red—in a row in the order given. But these colors are not sharply separated from one another so that, for example, we have blue on one side of a line and green on the other; they seem to melt into one another, passing through all the possible shades and tints of each color. Further, we have violet at

9

one end of the line and red at the other.   Now we know that
in the arts violet is produced by a proper blending of shades of
red with shades of blue.   Thus violet and indigo stand in the
series between red and blue, just as green stands between, and
is to be made by a blending of, blue and yellow, and orange
stands between yellow and red.   From these facts, we may
understand what scientists mean when they say that the solar
or light, spectrum is only one of a series of spectrums, each
beginning with violet or what corresponds to it in some other
form of force, and ending with red.

The Musical Scale.—Now, turning to the musical scale,
we have a fact precisely similar to the spectrum.   Here we
have another series of seven; sounds, which musicians name
A, B, C, D, E, F, G, so that whenever we run over the notes in
order, we find that every eighth note is the same sound as the
one we began with, only higher or lower, nearer the bass or
treble, as the case may be.   Now, we have beside the seven
"naturals," as musicians call them, five other notes called half
tones or "sharps" and "flats"—thus, "B" sharp is "A"
flat—and, furthermore, certain delicate instruments used by
experimenters have demonstrated the fact that each simple note,
as we suppose it to be, is in reality a bundle or collection of
many separate tones or sounds, differing in pitch and also in
loudness.

The Qualities of Sounds.—The pitch of each separate
note, as it affects the ear, is, in reality, only the pitch of the
gravest and loudest of the collection, and it is called the
"fundamental."   The other sounds are called the "overtones"
or "harmonics" and these, mingling with the fundamental,
give the timbre or quality of the note, although, when we speak
of timbre, we cannot by the ear separate the individual notes,
which combine to produce its character, pleasing or displeasing.
These overtones we may compare to the succession of shades by
which one color of the spectrum passes almost imperceptibly
into another, even while not interfering with the eye's sensation

of seven colors. In one kind of instrument, such as the piano, we have one set intensified, and in another, such as the violin, we have another set; and it is this fact that gives the characteristic difference between a piano and a violin note, although the "fundamentals" may be the same in both. Thus in a musical note the ear finds three things—the loudness of the note, its pitch or tone, and its timbre or quality.

Octaves of Force.—Having thus demonstrated that light and sound both follow one and the same law, scientists conclude that both are manifestations of the one force in nature. In other words they find that both are forces, manifestations of vibration in the air, in solid bodies, or in the finer "ether" which is believed to pervade both. As may be familiar to many, the difference between any two notes in one octave, and between the same notes in two different octaves, is one of vibration, slow or rapid. The ear can perceive as "one tone," a sound consisting of less than 32 vibrations per second, and can perceive a sound consisting of as many as 32,000 vibrations per second. This represents a difference of fifteen octaves. On the basis of the computed number of vibrations of the faintest ray of light, scientists have agreed to call the light area the "fiftieth octave," thus giving the difference between what the ear perceives and what the eye perceives, as thirty-five octaves of vibrations in regular scale of increasing intensity in multiples of 8. Moreover, we have no senses to perceive the effects to be derived from the vibrations in all this wide range of the unknown. Such marvelous modern discoveries as the Roentgen, or "X," rays may give us some idea of the facts arising under conditions we cannot "sense." For, here we have a power that can impress images on the sensitized film of the photographic plate, although we cannot perceive it as we can perceive light. Yet, as if it were in a way a stronger or subtler form of light, it can pass its rays through the living body of an animal, showing the outlines of the bones, or of foreign substances, and impress these on the sensitized film,

or make them visible to the eye, when viewed through the
"fluorescent" screen of calcium tungstate, devised by that
great genius, Thomas A. Edison.

Reflected Sound; Resonance.—Accepting the scientific theory that sound is a "mode" or form of force and
vibration, capable of producing on the ear a sensation we
describe as "hearing," we can readily understand that when we
speak or make other sounds with the vocal organs we do something more than expel air from the lungs. The air then used
is merely the motive power which works the vocal cords and
carries the vibrations they produce, high or low, loud or soft,
into the atmosphere, where they appear as sounds to affect the
ear. This process has a complete analogy in the experiments
of striking a tight string or a bell, having a given tone, and
observing that any other string, bell, or vessel capable of giving
forth the same note when struck, will respond, being set in
vibration by the sound waves that reach it. Thus, if we have
a room full of bells, all of the same tone, and strike one, all
will be set in vibration and begin to sound out the note belonging to them. If, however, there be bells in this collection that
give forth a different note from the one struck, they will not
respond. Such experiments prove that the true power of sound
is not in the mere motion of the air but in the kind or quality
of such motions; it is, in short, a question of the particular note
sounded in a given case, and of its characteristic number of
vibrations per second.

Sound Waves.—The way in which sound waves are set in
motion may be understood from what occurs when we throw a
stone into the water. Then we see, first a commotion in the
spot where the stone falls which at once becomes the center of
a series of rings that move from it, growing larger and larger,
as they are further away from the common center, and finally
ceasing altogether. But sound waves originate in an object
entirely surrounded with the atmosphere—we have heat and
light waves from outside, from the sun, but no sounds—hence

we must conclude that they are not "on one plane," like the waves on the surface of the water, but producing a series of spheres of force in the air, consisting of alternate moving layers and quiet layers, moving out from one center, which is the instrument making the sound. In these alternate active and

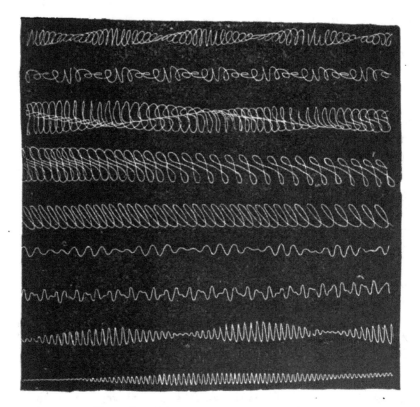

FIG. 10.—An illustration of the movements of sound "waves," or vibrations, as recorded by the phonautograph. These represent musical sounds; the lines produced by the human voice are extremely irregular.

inactive layers of air, we have a condition like the waves on water; first a "hump" and then a "trough."

Thus we have the fact of "sound waves," and their existence is due to the fact that air, unlike water, may be compressed. That is to say, we may by use of a pump put a mass of air many times the density of the atmosphere into a closed receptacle or "condenser." This is what is done when a bicycle tire is "blown up" with a hand pump, so that it becomes hard and firm. But air has also "resistance;" thus it is that one must use considerable strength in inflating a tire.

Movements of Sound Waves.—These two qualities of air unite in making sound waves possible. Take a closed tube, as shown in Fig. 11, in which a piston marked *P* can be worked back and forth. Suppose it is worked rapidly between the

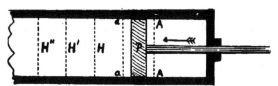

FIG. 10.—Illustrating the production of sound waves.

lines marked, *aa* and *AA*. When it moves in the direction of the arrow it slightly condenses the air in front of it, but because the air is at once compressible and resisting, this condensation does not affect the air in the whole length of the tube, but only that between *a* and *H*, which is then called the "condensed wave." It has been demonstrated that when the first "layer" of air between *a* and *H* comes to rest its motion is imparted to the first "layer" of air between *H* and H^1, and so on until the force exhausts itself.

Now, when the piston, *P*, is drawn in the opposite direction, that is from *a* to *A*, a vacuum, or air-vacancy, is produced behind it, causing the air to expand there to fill it up. Thus, we have, as it were, a back movement, which, like the forward movement, is transmitted from *a* to *H*, from *H*, to H^1, and so on, giving us the alternate active and inactive (or reactive) "layers." The forward movement of the piston from *A* to *a* shows exactly what is done when a sound of a note, a letter or a syllable comes out of the mouth in speaking. The back movement from *a* to *A* shows exactly what happens at the end of that particular sound.   Every sound is such a short and abrupt forward movement, due to the passage of a vocal sound through the mouth, when the tongue, palate and teeth are for a moment in some particular relative position, as will be explained later.

Vowels and Consonants.—All the sounds called "consonants" such as, B, C, D, F, G, etc., are short and explosive as any one can find out; that is a consonant sound may not be "prolonged," or sounded for more than an instant.   It is this abrupt ending that is like the back movement of the piston. But there is another class of letters, called "vowels," such as A, E, O, U, which may be made to sound as long as we wish, also certain "sonants," like M, N, L, R, S, Z, and Th.   But, if we will notice, we will find that each of these makes a "buzzing" or vibration that may be felt in the teeth, or perceived by the ear.   This "buzzing" is due to the fact that each vowel or sonant sound is made by a rapid series of "moves and stops" like the forward and back movements in the figure just described.   Thus in all sounds we have "hump" and "trough" movements, like waves on water.

Waves of a Moving String.—This fact may be beautifully illustrated by the experiment of striking a string drawn tight, so as to cause it to give out the note belonging to it.   While it is vibrating its sharp outlines are lost in the tremor that results, and in the middle it seems to swell, "belly out" as they say, into a hazy spindle of agitated matter.   The eye is incapable of following all the swift changes of the string's movements, and the result is a blur.   If now, we make "riders," by bending short strips of light paper in the middle, and place them on the vibrating string, one at a point one-third of the string's length from one end, and another midway between the two points, we will find that the first "rider" will remain on the string, being only slightly agitated, while the second will be thrown off like a stone out of a sling.   Or, if we place sharpened blocks beneath at points one-third and two-thirds the length of the string, we will find that they do not materially interfere with the vibration or pitch of the string   These facts prove that there are alternate series of activity and inactivity in any moving string, the former being called the "nodes" and the latter, "loops;" thus we have a regular wave motion, humps and troughs.

FIG. 12.—Diagram of Leon Scott's Phonautograph.

How Sounds May be " Seen."—The wave motions of sound have been still further analyzed and recorded by delicate and ingenious instruments, such as Leon Scott's "phonauto-graph" (*i. e.,* "sound-self-writer") and Koenig's apparatus for producing what are known as manometric (*i. e.,* "thinness-measuring") flames.   The former consists of a sounding-chamber or barrel, open at one end to admit the sound and closed at the other by a brass tube with a flexible membrane stretched at its extremity.   To this membrane is attached a hog's bristle which is able to trace the lines of vibration affecting the membrane on soot-covered paper wound about a cylinder revolved by a hand-crank.   The tracings shown in Fig. 10 illustrate some of the common forms of sound-autographs, thus produced. Some of them resemble the lines traced by the sphygmograph of the pulsations of the blood.

Manometric Flames.—Koenig's apparatus is equally ingenious and effective.   It consists of a metal "capsule" or box, divided into two chambers by an elastic diaphragm. At one end is an orifice for the admission of sounds from the voice or an instrument, and at the other is an ordinary gas-burner, gas for which is admitted to the chamber in front of the diaphragm. If the gas be lighted and its supply be agitated by the sound waves striking against the diaphragm the flame will undergo perceptible changes in strength and steadiness. These may be seen to advantage if we take a four sided box with mirrors on

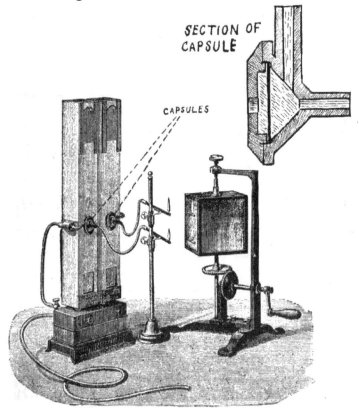

SECTION OF CAPSULE

CAPSULES

FIG. 13.—Koenig's Apparatus for producing Manometric Flames.

every side and turn it rapidly in front of the flame. The resulting "image" will differ according to the strength, timbre and other characteristics of the sound, giving such figures as may be seen in Fig. 14.

The Human Voice.—Wonderful as are the instruments for producing such effects the vocal organs and ear of a human being are even more remarkable in their complexity, adjustment and range of powers. The mechanism for producing articulate speech consists of two important parts; the sound-producer and the sound-shaper. Both are peculiar to man in the range of sound possibilities and in the power of shaping. Other animals can make their distinctive calls and cries, expressing their emotions and needs, and such noises are to a certain extent shaped and definite sounds, besides partaking of musical qualities—timbre and volume—but in none of them is there anything approaching to the human power of articulation.

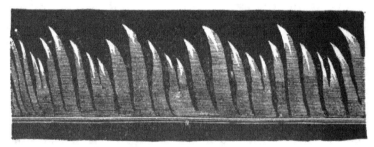

Fig. 14.—Manometric Flames produced by Koenig's apparatus. These illustrate a blending of two distinct musical notes.

The Vocal Cords.—The sound-producer in man is the larynx, and its action consists in forcing air from the lungs through a slit between two membranes, known as the "vocal cords." By means of a number of delicate muscles these may be tightened or relaxed, so as to produce high or low notes as the case may demand. In the full, deep sounds, known to musicians as "chest notes," the entire membrane is found to vibrate, but in the high notes and in the "falsetto" only the edges are used. This fact will be seen to depend on the same principle as that by which the strings of a guitar give forth differing notes, according as they are shortened by the pressure of the finger on this or that fret. The characteristic difference between the voice of a man and of a woman is due to the fact that in the former the vocal cords are larger and thicker, and the larynx longer, than in the latter; hence able to vibrate less rapidly. The action of the vocal cords, may be readily compared to the action of the lips in whistling. An attentive study with a looking glass will reveal the fact that the lips open more in the production of low notes and contract, sometimes almost to closing, in the production of the high ones.

Powers of the Voice.—The vocal chords, together with the mouth and the several resonant cavities opening into it, are able to produce a wide range and variety of sounds. The average reach of the human voice is about two octaves—some voices reach as far as three, but few beyond that—and the power of varying the timbre, to permit the imitation of various musical instruments, animal calls and other sounds and noises seems almost unlimited. That the ear is capable of perceiving the wide variety of tones and shapes argues that it is a most delicate instrument.

The Organs of Hearing.—Fig. 15 gives a good general view of the series of organs, by which sound waves entering the "meatus," or opening, of the outer ear are transformed from atmospheric vibrations to nerve pulses that can affect the brain with definite and finely separated impressions of sound. The process by which one form of force is thus changed to another may be compared to the transformation of heat to motion, as through the steam engine, or of motion to electricity, as in the dynamo electric generator. The mechanism of the ear thus combines the functions of both separator and transformer, as those terms are used in mechanical science. The essential parts, as shown, are: 1, the tympanum, or "drum," a membraneous diaphragm stretched across the inner end of the external canal; 2, a series or jointure of three minute bones, or "ossicles," called respectively the *malleus*, or hammer, the *incus*, or anvil, and the *stapes*, or stirrup, the first pressing against the inner side of the drum, and the third, against another membrane closing the passage to the inner ear; 3, the inner ear, cor

sisting of a marvelous series of canals in the bony substance of
the skull, known as the "labyrinth" and the "cochlea." (snail-
shell); 4, the auditory nerve attached to the nerve fibers filling
the former and also running into the latter through its turns, or
"convolutions."

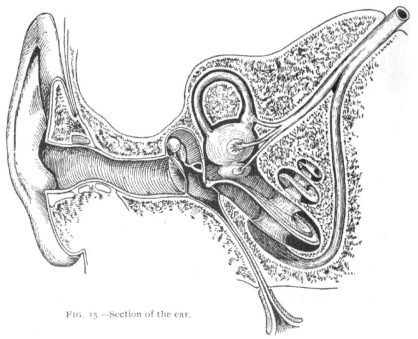

FIG. 15.—Section of the ear.

The "Inner Ear."—The inner membrane of the cochlea is
lined with some 3,000 minute fibres, which are the ends of the
auditory nerve. They are known as "Corti's fibres," and each
seems to be tuned to some particular note or tone, which,
conveyed into the ear by the sound waves from the outer air,
can cause it to vibrate in unison, just as the bells tuned to the
same note. Each fibre is unresponsive to other notes, just as is
a string or a bell, and thus we may understand how that some
disease among them will make a person "tone-deaf," or unable
to distinguish one note or tune from another. Some persons
seem to be so affected that they cannot hear certain notes, which
are then said to be "out of their register."

How do We Hear?—Exactly how the nerve fibres of the ear are able to respond not only to notes of different tone, but also to sounds of different "shape"—articulate sounds—we have no satisfactory idea. No more is it plain how the minute dents traced by vocal stress in the tin-foil or wax on the revolving cylinder of the phonograph are capable of reproducing the same sounds, words, timbre and tones when the process is reversed. We can derive some idea of the process by which an electric current, alternately interrupted and closed, by a series of "dots" and "dashes"—shorter and longer connections, as in the Morse telegraphic alphabet—can convey intelligible mes-

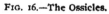

FIG. 16.—The Ossicles.   FIG. 17.—Cast of the Labyrinth.

sages from one end of a line to the other. But the principle by which the telephone transmits articulate sounds by this same process of interrupted currents is, and must continue, a mystery in spite of all our learned theories and attempts at explanation.

The "Sound Shaper."—As we have seen the sounds of the human voice are produced in the larynx, by means of the organs known as the vocal cords. But by these we have only simple sounds or tones, such as the same amount of air would produce from a properly constructed reed, organ pipe, or trumpet, or such as would result from scraping or picking a taut string. The mechanism for shaping, articulating speech, is found in the lips, tongue, teeth and palate. In obedience to the dictates of the will, these assume various relations, so that the sounds passing through them at no matter what pitch or timbre, invariably issue from the mouth in the forms peculiar to human speech. The lowest human languages have been described as a "series of grunts, clicks and hisses," but the most highly organized forms of language present a wide range of true speaking sounds, from the pure gutturals, such as hard G and K, to the pure labials (lip letters), B, P and M. Some languages, like the Sanskrit, the "sacred tongue of ancient India," recognize a variety of sounds we fail to discriminate in English, except in theory, having four sounds for N, two each for D and T, three for S, and using an R and an L vowel, and practically also an M and an N vowel.

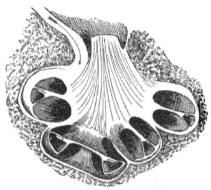

FIG. 18.—Section of the Cochlea.

While, in the production of articulate sounds the lips and teeth have a very important part, by far the greatest work is accomplished by the tongue, which changes its shape and its position as regards the teeth and palate with remarkable rapidity. As may be readily understood, its office is to so modify each sound passing through the mouth as to vary its character by varying, as we might say, the resistance with relation to the two resonants, or sounding-boards, the palate (roof of the mouth) and teeth. We may assume that such reflections of the sound set up a number of vibrations in different directions and at different rates, instead of the simple and direct waves of the pure sound. The process is perfectly plain to any one, but the reason therefor may not be fully explained. We may, however, find proof of the fact that some such variation of resistance will literally change the character of a note, not in respect to its tone or timbre, but in its shape, by certain experiments, as illustrated in the following figure:

Voice Pictures.—Fig. 19 illustrates one of the many effects on small heaps of light sand or other fine powder, moistened to a pasty consistency, and placed on a flexible diaphragm stretched over the mouth of a tin can, into which sounds from the voice are introduced through a pipe like the spout of a teakettle. A wide variety of such shapes has been made, according to the shape of the vessel covered by the diaphragm, and according to the note sounded into it. The explanation is simple. The figures are not mere visual forms corresponding to the vibrations made by the note producing them, but result from a variety of forces working together. Thus, the sound waves made within the closed bowl are reflected again and again from its sides. The mass of air enclosed has one rate of vibration, the stretched membrane another, and the layer of paste a third. These combine to give the elements of reflected sound and resistance, thus making shapes to be seen rather than shapes to be heard, as in the case of speech.

FIG. 19.—The Eidophone and "voice picture."

We have thus seen that sound consists in vibrations set up in the air, in solids and in liquids, and is, so far as we can understand its nature, a kind or degree of motion among the particles supposed to constitute all material things. Sounds may have four qualities—volume, tone, timbre and shape. The last is found best in the human voice. Moreover, all may be transmitted as far as the sound will go; they all combine to make up a single impression on the brain through the ear. They may also together make impressions that will similarly vary an electric current. On this fact rests the theory of the electric telephone.

As is obvious to any observer, there is a marked limit to the power of sound to carry to a distance in the atmosphere, or other medium, by the simple force of its own vibrations. The sound of a church bell may be heard for several miles, under proper conditions, but if the wind is in the wrong direction it may not be heard for more than half a mile, and sometimes less. From this illustration we may understand that the moving mass of air we call the wind carries the sound waves along with it, sometimes miles beyond the point where it can usually be heard least distinctly. Thus it is that two kinds of

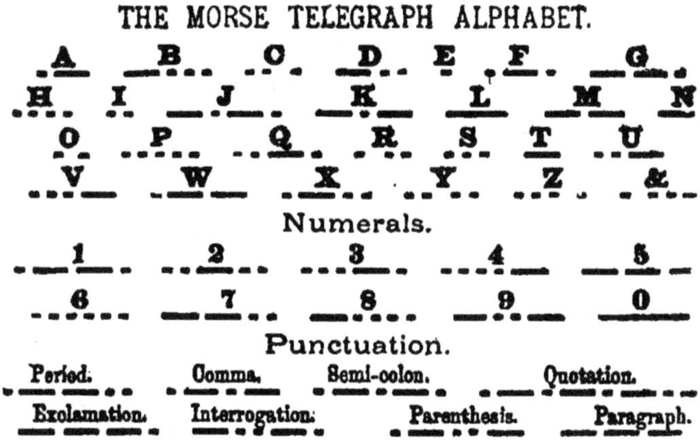

THE MORSE TELEGRAPH ALPHABET.

Fig. 20.—The Morse Telegraph Alphabet, illustrating the method of transmitting intelligence by "making" and "breaking" an electric current. The dots are short "makes"; the lines, longer pauses.

motion, that of the mass of the air, and that of the molecules composing a part of it, which we call sound waves, can so blend that they move along side by side or in the midst of each other. In other words, the range of the waves of sound is greater in one direction when the air is moving thither than when the air is still. We may illustrate this by the familiar fact of a rapidly flowing stream of water on which we may observe ripples moving across from side to side, caused by the resistance of the banks and channel, or occasional obstacles, like rocks or trees. We see the transverse (crosswise) vibrations carried along by a forward-moving current. In some such way the waves of sound are conveyed along a current of air, although, like the cross-waves on the water, they must sooner or later exhaust their force and disappear.

Just as sound may be carried on the moving air, it may be also conveyed along any direct or confined channel, like an old-fashioned speaking tube, formerly so common in houses and business offices. The tube acts to prevent the sound waves from spreading through the air, and hence can carry distinct words to a greater distance than they could be plainly heard without it. The ordinary acoustic telephone, or "lovers' telegraph," previously described, is another example of the preservation and conduction of sound waves along a definite line or conduit. The vibrations being communicated directly to the taut string are carried along its length without suffering speedy dissolution by atmospheric resistance.

To make the telephone a thoroughly practical machine, however, it is essential that we have some agency that will at once carry the impulses of the sound waves to long distances, as does the wind, and also keep them confined to a narrow and definite channel, so they may reach the intended destination, and no other, in as strong and distinct a condition as possible. This agent is found in the electric current, which is able to take up and transmit the timbre, pitch and shape of sounds, so that they may be reproduced at the distant end of the line.

CHAPTER THREE.

A BRIEF SURVEY OF THE PRINCIPLES OF ELECTRICITY.

Electricity Defined.—Electricity may be defined as a "powerful physical agent which manifests itself in attractions, repulsions, heat, light, commotions in matter, chemical decomposition, and other physical happenings." As to its exact nature, there are many theories, much debate, and nothing definite. It has been called a "fluid," but so were light and heat in former times. Whether it be simply a mode of force or not, two things are certain: 1. It never appears apart from some physical change in matter. 2. It always affects changes in matter, as do heat and chemical activity, by alterations in its quality or condition. For all practical purposes we may speak of electrical force as a form of energy, having its peculiar characteristics, as has any other form; and the fact that we cannot fully comprehend or reconcile all the facts we know of it is scarcely remarkable, when we remember that human knowledge never reaches beyond the range of facts experienced, nor touches final principles.

Varieties of Electrical Force and Source.—With these points understood, we may say that electrical force is always a product; that is, it always results from some condition or activity in matter, which exists when it appears. Anything that produces electricity is called an electrical "source." According to effects, we may define two kinds of electricity: static, or stored, electricity, which is discharged in the form of a shock or spark, as from a Leyden jar or the clouds in lightning; and current, or dynamic, electricity, which flows constantly between the two poles of a "circuit," along the line of a wire or other "conductor." According to source, there are three kinds of electricity: frictional, such as is produced by rubbing, as with

26

one of the several kinds of electro-contact machines, or the simpler electrophorus; magneto-electricity, such as is produced by the modern dynamo-electrical generators, commonly called "dynamos;" and voltaic or galvanic, such as is produced in a chemical battery cell by the decomposition of a fluid, called the "electrolyte" by a current moving between two plates of dissimilar metals, known as the "electrodes." On account of many resemblances between the action of a current of electricity and those of a bar magnet, some writers make still another classification, calling magnetism "electricity in rotation."

Electrical Currents.—The fundamental truth recognized in all the modern applications of electricity is that there are not many kinds of electricity, but one only. If we take an electrophorus, a disc of resinous substance in which electrical excitation may be produced by rubbing with a silk handkerchief, and bring it into contact with any substance in a "negative" state the result will be a "shock," or discharge, like the transference of stored electricity from one cloud to another in the form of lightning in a thunder storm. This seems to be "the whole of it," as we might say. If, now, we take two plates of dissimilar substances, say copper and zinc, and immerse them in a solution of sulphuric acid, we will find that electrical activity of a continuous variety will be set up the moment the two plates are connected by a wire outside the solution. That the cause is the same in both instances is shown by the fact that effects peculiar to all phases of electricity follow; the production of electric sparks and the like. Further, although the results attending the shock and the current seem to distinguish them they are in reality the same thing. The shock manifests the fact that a substance holding a large store of electricity will "electrify," as the term is, a substance having a lesser store. The process is momentary because the source is not constantly supplying fresh energy. In the case of the current the source is constant; hence the movement from the high to the low potential is continuous, forming a current. This fact was first uncovered by

Galvani in 1786, when he observed the twitching of a frog's legs, hung on a copper hook and accidentally brought into contact with an iron bar. He supposed that the effect was due to the presence of "animal electricity" in the legs, but it was shown by Volta that it is due to the contact of dissimilar metals, differing in potential, which is to say, power to contain or give off electrical energy.

The Electric Circuit.—In order to have an electric current, we must have a "circuit," which is to say, a closed path of wire or other conducting substance. In the voltaic cell the

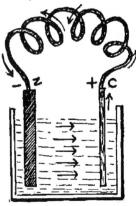

circuit is formed by connecting the copper electrode with the zinc electrode, so that a constant flow of electrical energy is set up, through the solution in which they are immersed, from the zinc to the copper, and, on leaving the solution, from the copper back again to the zinc. An electric circuit is thus, as the term implies, a means for "going around." We have a very good illustration of the circuit of an electric current in the conditions which create a draught of air. We cannot have a true draught in a closed room with only one outlet, unless there be a considerable difference in temperature between that room and the air

Fig. 21.—Diagram of typical Galvanic Cell, showing circuit "made" or "closed." If the two electrodes, Z and C, were not connected by the wire, no current would flow, and the circuit would be said to be "broken" or "open."

outside. This creates a circulation, until the temperatures become the same. To have a true draught, we must have an inlet for the air to enter the room, and an outlet, to enable it to escape into the atmosphere again. Thus, a fire in a stove creates a draught, that is draws air through the draught holes, so long as the chimney is open to allow the air to flow out freely. But so soon as we close the draught hole or shut off the connection with the outside air through the chimney, by closing the damper, we interfere with the circuit of the air current, and the draught ceases. Thus, to continue the comparison, the fire in the furnace or stove, consuming coal or wood, is to the air current what the battery is to the electric current. It has the power to set up a form of activity. In the stove the activity is produced by a condition called "heat;" in the battery, by a condition called "electricity." Both require a circuit—a going-in and a coming-back.

Electromotive Force, E M F.—When a current is given off by a chemical or magnetic battery, it is a popular error to describe it as electricity. To be correct, however, it is nothing of the kind, but the resultant of a produced electrical condition, and is defined as electromotive force. This force is usually indicated by its three initial letters; thus, **E M F**. This distinction is made because there is a marked difference between the force which travels on the wire and produces effects and the conditions which gave it birth. Hence we must use separate terms to indicate the two.

The Voltaic Cell.—The simplest form of the voltaic cell is, as we have seen, an electrode of copper and an electrode of zinc, immersed in a weak solution of sulphuric acid, and connected at the upper end by a wire forming a circuit in which is placed any body or contrivance to be affected by the current. Before this circuit is formed or "closed," as the term is, there is no action whatever, beyond a slight "disengagement," or separation, of hydrogen gas on the face of the zinc plate. So soon soever as the electrodes are connected so as to form a circuit, the action is removed from the face of the zinc to the face of the copper, where the active chemical condition is to be observed. The continuance of the current is characterized by a wasting of the zinc electrode, due to the combination of its particles with elements set free by the decomposition of the electrolyte into the constituent elements. As has been well said, this fact proves the cell to be a "kind of chemical furnace in which the fuel is zinc, or the negative electrode."

Activity in a Voltaic Cell.—The current thus formed is supposed to pass out through the wire from the positive electrode, and to complete the circuit by returning and passing in through the negative electrode, and thence through the electrolyte back again to the positive electrode. The process of electrolysis, or electrical breaking-up of a substance, as it takes place in a chemical battery cell, is indicated by the diagram shown in Fig. 22. Here we have three rows of figures, indicating so many molecules constituting the liquid surrounding the electrodes. Each is made up of two elements, as indicated by the light and shaded portions. As in the simplest form of chemical cell this liquid is diluted sulphuric acid, whose components are two parts of hydrogen, one part of sulphur and four parts of oxygen ($H_2 SO_4$), the light end of each molecule represents H_2, let us say, and the dark portion SO_4.

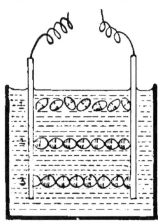

FIG. 22.—Diagram of Grotius' theory of the galvanic cell. Row 1 shows position of molecules when circuit is open; 2, at the moment of closing; 3, when current is flowing.

Grotius' Theory of the Voltaic Cell.—Now, according to Grotius' theory, which this figure illustrates, the upper row of molecules indicates the condition before the electric circuit is closed, when there is no special order in the arrangement. The second row shows what takes place from the moment the current begins to flow along the circuit. Now all the molecules arrange themselves in regular order with their unlike "poles" or ends in contact. The new arrangement is due to the fact that each one is "polarized," or has acquired magnetic qualities. In addition to this "polarization," all the molecules are supposed to move in regular order, in lines, from the zinc electrode to the copper, and back again. In course of this "rotation," or traveling around, we find that the sulphur and oxygen unite with the zinc, causing it to waste, and the hydrogen, thus liberated, unites, with the SO⁴ of the molecule just ahead; that in turn giving off its SO⁴ to the one in front of it until the copper plate is reached, and there we find it is given off to unite again with the liquid and perform the same rotation, until all the SO⁴ has united with and consumed the zinc plate. This is, in brief, the process by which the electric current is produced in a voltaic cell.

Magnetic Lines of Force.—The current from an electrical cell, as it passes steadily from the point of positive, or high, potential to that of negative, or low, potential, finds analogy in

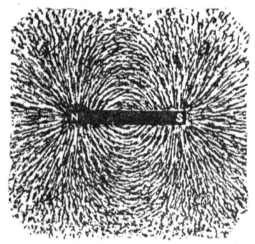

FIG. 23.—Lines of force in a bar magnet. Shown by dusting iron filings over the "field."

what are called the "lines of forces" in a magnetic "field." If we take a permanent bar magnet and place it near a quantity of iron filings we will find that these will be attracted to both poles, and will adhere there with considerable force. But toward the center of the bar we will see very few. This indicates that there is some system in the magnetic power of attraction, and that it is not the same at all points.

If now we place over the bar magnet a plate of glass or a sheet of stiff paper, and on it dust a quantity of iron filings, we will find that they arrange themselves as shown in Fig. 23; lines of filings starting out from either pole, and at points midway between, which seem to describe arcs (arches) or parts of circles, over the bar, meeting in the middle of the "field."

The lines thus described by the filings are supposed to represent the lines of force, or the direction of the current, as it leaves the north pole and re-enters at the south. This fact is illustrated in Fig. 25, where the lines of force are indicated by arrows supposed to enter and leave the magnet as lines of force.

Fig. 27 further illustrates the theory of the magnet, showing how that filings arrange themselves as rays, or straight lines starting from a common center, when scattered on a sheet of paper or glass placed over either end, or pole, of a magnet. In this figure we have the end view of the semi-circles of force shown in Fig. 23.

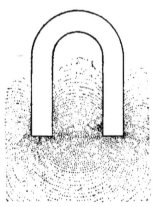

FIG. 24.—Lines of force of a "horseshoe" magnet.

Magnets and Magneto-Electricity.—On the fact of these lines of force in a magnetic field is based the theory of the dynamo-electric generator, commonly called the "dynamo." In this machine the current is obtained by producing mechanical excitation across the lines of force in an electro-magnet. It may be well to explain here that there are two kinds of magnets. The first is the electro-magnet, consisting of a soft iron bar, around which a silk-covered wire is wound a large number of times, like thread on a spool. When an electric current is passed through the wire the iron bar receives all the qualities of a magnet—power to attract iron, etc.—which continues so long as the current is passed through the wire, and ceases when it is withdrawn. The second kind of magnet is the permanent magnet, consisting of a bar of hardened steel. The magnetic qualities are given to it in exactly the same manner—passing a current through the coil or "helix" of wire—but are retained by it for an indefinite time after the current ceases. The horseshoe magnets used in telephone generators and receivers are usually made in another way—stroking the hardened steel bar across the poles of a very powerful electro-magnet. Either form of magnet may be used as the basis of a magneto-electrical generator, but it has been found most suitable to employ electro-magnets in commercial dynamos of large power.

Although the construction and use of dynamos form a distinct branch of electrical industry, the principles involved are applied not only in the construction of the telephone and the magneto-generator, but are also becoming increasingly important in other branches of telephony, such as the modern central energy exchanges. Fig. 28 shows the simplest possible form of dynamo. Here, *N* and *S* rep-

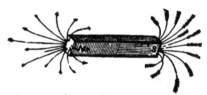

FIG. 25.—Diagram illustrating the theory of magnetic "circulation" along the lines of force.

resent the two poles of a " horseshoe," or curved-bar, magnet, between which lines of magnetic force flow, as shown by the dotted lines. Between these poles the wire loop, *C C*, revolves on the bar or spindle, *A A*, thus cutting across the lines of force. This process makes a current which flows out and back again on the circuit wire, *E*, which is attached to the "brushes," *B¹* and *B²*, so-called because they *brush* against the revolving "drums," as shown in the cut.

How the Dynamo Works.—In this figure the loop of wire would be called the "armature," and the greatest difference between this simple machine and the powerful dynamos used to produce currents to move street

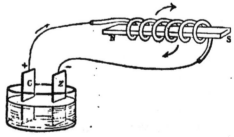

FIG. 26.—Diagram of the usual method of producing magnets.

cars and supply electric lights is in the construction of the armature. In the latter, instead of a simple loop of wire, there are a great number of such loops or of coils, wound on a "drum," or "ring," and running at right angles to the magnetic lines of force. The rule is that, electromotive force (E M F) generated by any dynamo, is in proportion to the number of turns of wire about the "core," or frame-piece, of the rotating armature; and within certain limits, to the speed with which the armature is revolved. Owing to this arrangement, the lines of

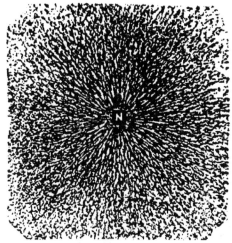

FIG. 27.—Lines of magnetic force at the pole of a bar magnet.

force are greatly deflected, as shown in the dotted lines in Fig. 29, and by the turning of the armature a very intense excitement of the "field" is created. From this excitement a current results, whose strength for work is in proportion to terms of the rule quoted above.

The effect produced by the revolving of the armature of a dynamo is that the armature itself is transformed into an electro-magnet, having two north poles and two south poles, points for the exit and entrance of the current, respectively, as is shown in Fig. 31. These poles alternately exert an attracting and repelling force on the magnetic field, and hence continually distort the magnetic lines of force, as shown in Fig. 32. The result is a constant shifting of the neutral points, and a continual series of alternate attractions and repulsions at the active points. These rapid changes, demanding a constant readjustment of the magnetic lines of force, transform the magnetic movements among the molecules into electrical energy, which emerges from the machine in the form of a current.

Magnetic Properties of "Live" Wires.—Not only is it true that the polarity of magnets gives a good idea of the theo-

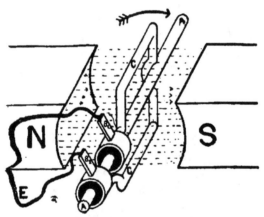

Fig. 28.—Simplest possible form of dynamo. *N* and *S* are the poles; *C* is the armature; *B B* the brushes and *E* the line wire.

retical passage of an electrical current from a point of higher positive, potential, to one of low, or negative, potential, through the air or any suitable substance, such as is known as an electrical "conductor," but it has also been observed that a wire on which an electrical current is active shows signs of magnetic activity. If we pass a wire through a hole in a card or piece of paper and then connect it into the circuit of a battery or dynamo, causing a current to flow on it, we will find that iron filings will arrange themselves about as is shown in Fig. 35, the wire being the center of a number of circles, made by the iron filings arranging themselves as they are impelled by the force around it. This fact shows that each wire carrying a current is surrounded by a complete whirl of magnetic force and influence, as shown in Fig. 36. Thus, we may say that every current-carrying wire has an electrical effect along its length, and a magnetic effect at right angles to its length.

Electrical Conductors; Non-Conductors.—Although the whole subject of the generation and use of the electrical current is filled with mysteries and unexplained points, we may treat it as a force on account of the fact that it always follows along the "line of least resistance," as the term is; that is to say, it always avoids the path that is most filled with obstacles to its flow, and chooses the one that is most easy. From the observations on this fact, scientists have found that all substances are divided into two classes: conductors, along which the electrical current flows with ease, transmitting its effects from one end of the line to the other; and non-conductors, or insulators, in which it meets with difficulties, or is not able to flow at all.

The following substances are good conductors, the best being first named and the less good following in order:

Substance.	Relative Conductivity. At Zero, Centigrade.
Silver	100.00
Copper	99.90
Gold	80.00
Sodium.	37.40
Aluminum.	34.00
Zinc	29.00
Cadmium	23.70
Brass	22.00
Potassium	20.80
Platinum	18.00
Iron	16.80
Tin	13.10
Lead	8.30
German silver	7.70
Antimony	4 60
Mercury	1.60
Bismuth	1.20
Graphite	0.07

The conductivity of all these substances having been computed at a common temperature, we have as correct an idea of their relative qualities as is possible, on a common standard. It has been found, however, that a number of conducting substances, including water and most acids, increase in power to conduct electricity with an increase in temperature.

In addition to the above list there are a few substances, which, while they are capable of conducting electrical currents,

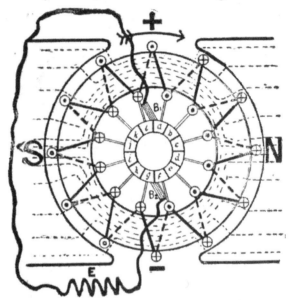

FIG. 29.—Diagram of a ring armature dynamo, showing the deflection of the lines of force.

vary in the capacity according to conditions. They are, briefly:

Animal bodies, cotton, dry wood, marble, paper.

The substances that will not conduct the electrical current, and will interfere with its transmission are:

Oils, porcelain. wool, silk, resin, gutta-percha, shellac, ebonite, paraffine, glass.

They hold the same relation to the electrical current that any solid body does to the flow of water—they check or dam it. On account of this property they are extensively used in all the branches of electrical industry where it is desirable to confine the current to definite limits, as to a wire, or to prevent it from spreading beyond the particular machine it is desired to "charge." They are thus called "insulators" (Latin, *insula*, an island; *insulatus*, made into an island). For purposes of thus confining, or insulating, an electrical current, telegraph and telephone wires are attached to poles by the familiar glass caps, and the wires used about electrical machinery, or where

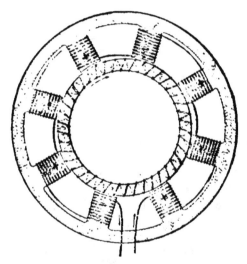

FIG. 30.—Diagram showing the lines of force of an eight-pole dynamo, as they circulate between the pole pieces and the armature. Direction of currents shown by arrows.

likely to come in contact with wood or other combustible materials, are covered with silk-windings or india-rubber.

The Strength of Magnetic Force.—While insulation of a wire will interfere with the spread of the current, it has no effect on the magnetic forces, which, as we have seen, accompany it. Thus it is that when a great length of silk-covered wire is wound about a bar of hard steel, as thread is wound on a spool, it is able to make a permanent effect upon it; to transform it into a permanent magnet. Although in this process the wire is so thoroughly insulated that the electrical current follows along its entire length—not striking across, and thus shortening the circuit, as would be the case were there no silk-winding—we find that there is an agency that is able to set

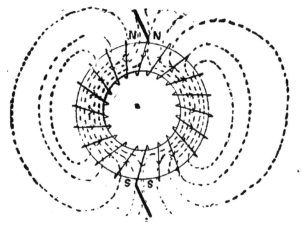

FIG. 31.—Diagram illustrating the polarization of the armature of a two-pole dynamo.

up the circulation of the lines of force that we have seen pertain to all magnets. And this is a result that would not follow if a voltaic current were run directly through the bar of steel.

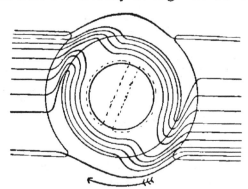

FIG. 32.—Diagram illustrating the distortion of the magnetic field in a two-pole dynamo by the revolution of the armature.

Electrical Induction.—In this occurrence we are introduced to another department of electrical activity. As the effects of a current carried from one end to another of a wire are said to

be due to "conduction" (a leading-along-with), so the effects
that are thrown off at every point as it moves along—these are
the magnetic effects—are attributed to an agency called "induction" (a leading into.)

That it is proper to attribute the effects of induction to
magnetic influence is evidenced by the fact that there are no
magnetic insulators or cut offs. A magnet will attract iron, as
we have seen, through a pane of glass, which would effectually
cut off a current of electricity. Consequently, it is able to produce its effects in spite of the silk-windings on a magnetizing
wire.

Currents in Solenoids.—If we take an insulated wire and
coil it, as shown in Fig. 38, like an ordinary wire spring, with
one end passing through the coil, we will find that the action

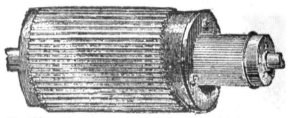

FIG. 33.—The Edison Bar Armature a typical commercial dynamo armature. The
smaller cylinder at the right end is the commutator, constructed with alternating strips of
conducting and non-conducting materials, on which rest the brushes that gather up the
current.

of a current passed through the coils and out along the line of
the straight wire is exactly equivalent to that of a series of
parallel currents running in that direction. This contrivance is
called a "solenoid," and it has been found that when two of
them charged with currents are brought together, the action
will be the same as if they were two magnets; the ends where
currents enter or leave will repel each other and attract the
opposite extreme. This shows that the electric current will act
in magnetic effects between two insulated circuits.

Magnetic Induction.—Again, if we have a circuit of wire,
enclosing a "galvanometer" (a machine used to indicate the

strength of a current), as shown in Fig. 38, and bring an ordinary bar magnet near it, we will see by the galvanometer that a momentary current is produced in the wire. The needle of the galvanometer then resumes the position it occupied before the production of the induced current. On the withdrawal of the magnet another current is produced, showing similar strength, but in the opposite direction.

Current Induction.—Fig. 40 shows the production of an induced current in the wire circuit B, carrying a galvanometer,

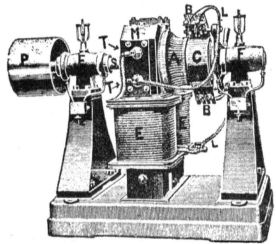

FIG. 34.—A complete dynamo. $M M$, pole pieces bored to admit rotating armature; $E E$, coil; A, armature; C, commutator; $R B$, brushes; $L L$, leads for the current; $T T$, terminals to line; S, spindle; $F F$, bearings; P, pulley.

by bringing into its vicinity the circuit, A, charged by a battery. So soon as this circuit is closed by the push key, we will observe that the needle of the galvanometer dial indicates a momentary current in the other circuit, only running in opposite direction. So long as the current A is kept flowing no other effect will be observed, but if the circuit again be opened, we will see by the galvanometer that a weak momentary current is induced in B, moving in the same direction as that in A.

Rules Governing Current Induction.—The principle of current, or voltaic, induction, thus described, is a consideration of the utmost importance, not only in telephony but in every other branch of electrical science. The following points, there-fore, are important:

1. Increasing the strength of the current in *A* increases the strength of the current in *B*.

2. Decreasing, or opening, the current in *A* decreases the strength of the current in *B*, also causing it to flow in the same direction as the current in *A*.

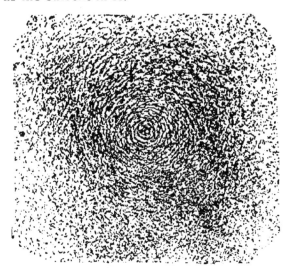

FIG. 35.—Concentric circles of magnetic force around a " live," or current-bearing, wire.

3. If we move the current-carrying wire, *A*, nearer to *B*, we produce a strong current in the opposite direction; if we move it farther from *B*, we induce a weaker current in the same direction.

4. If the wire used in the circuit *A* is thick, and that used in *B* is thinner, the current induced in *B* will show a greater electromotive force than that in *A*. Conversely, if the wire used in *A* be thinner than that used in *B*, the induced current will show a lower electromotive force than that in *A*.

Electrical Resistance.—The last rule holds true because of the difference between the two wires in a quality known as

" resistance." On this point there is considerable analogy between the action of a current of water and a current of electrical
energy. Fig. 41 shows two pipes, one large and one small, opening into the front and back of a cylinder. If, now, the pipes
be filled with water, and the piston be moved backward and·
forward, considerable resistance will be experienced in moving
the masses in the two pipes. That there is a difference on this

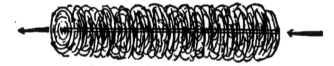

FIG. 36.—Illustration of the whirl of magnetic force around a current-bearing wire.

point between the two may be found by closing a stop-cock,
attached to either pipe, and merely moving the water in the
other. By this means we will find that we experience less resist-

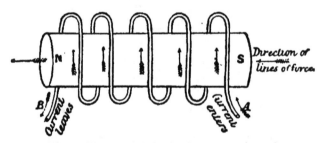

FIG. 37.—Diagram illustrating induction between a wire and a magnet.

ance in the short wide pipe than in the long narrow one,
although both may be equally smooth in bore and contain precisely the same amount of water.

The same is true of two electrical circuits, one shorter and
of thick wire and one longer and of thin wire. A given
strength of current passed through both will show a greater
resistance in the longer and thinner wire than in the shorter
and thicker wire, although both wires may weigh exactly the
same and contain the same amount of metal.

Laws of Electrical Resistance.—These facts give us the following rules, which are of importance in telephone lines and apparatus :

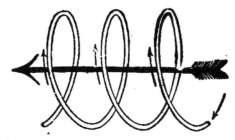

Fig. 38.—Diagram illustrating the principle of the solenoid and of magneto-voltaic induction.

1. Doubling the cross-section of a wire halves the resistance. Halving the cross-section doubles the resistance.

2. Doubling the length of a wire doubles the resistance. Halving the length of the wire halves the resistance.

3. To sum up: The resistance is measured in inverse ratio to the cross-section of the wire, and in direct ratio to the length of the wire. A thorough understanding of this principle enables the construction of electrical instruments of great delicacy and exactitude of effect.

Fig 39.—Usual form of galvanometer.

The Induction Coil.—One of the most useful and familiar contrivances based on the variation of resistance between two sizes of wire is the "induction coil." Its construction is as follows: A coil of insulated wire is passed around a bobbin over a core, which is either a bar of soft iron or a bundle of fine iron wires. In this we have the parts of an electro-magnet completely reproduced, and a current passed through the wire will induce magnetic properties in the core so long as it endures. Around the first coil of insulated wire we now wind another, only of much finer cross-section, and, generally, of greater length. If, now, we pass a current through the first coil of thick wire, which is called the "primary" coil, a momentary current will be induced in the second winding, which is called the "secondary" coil. Alternate closings and openings of the primary circuit cause a series of momentary currents to flow through the secondary circuit.

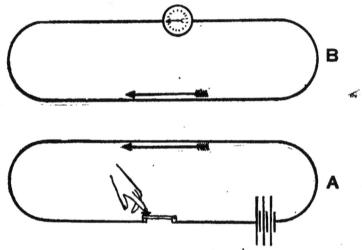

FIG. 40.—Diagram illustrating induction between two circuits; one energized by a battery, the other bearing a galvanometer.

Theory of the Induction Coil.—In the induction coil we have two kinds of induction—magnetic and voltaic—perfectly exhibited. The principal difference in effect is that the former continues as long as the current endures, while the latter is momentary, indicating the moments of closing and opening the circuit of the primary coil. It is possible that we have an explanation of the difference in the fact that the momentary currents in the secondary coil indicate the successive adjustments and readjustments of the lines of magnetic force, as the core is alternately magnetized and demagnetized. The effects of simple voltaic induction, seen between two wires, as previously described, is probably due to the influence of the magnetic area of the "live," or current-bearing, wire on a closed, but currentless circuit, in which it produces magnetic properties; that is, effects a molecular readjustment in the "dead" wire.

Specific and Comparative Resistance. — Having thus explained the properties and effects of currents, both conducted and inducted, it will be necessary, for a further understanding of the subject, to discuss the conditions under which electrical energy is passed along wires, and how it is measured. As we have seen, there is such a thing as resistance to the transmitted current in a wire,

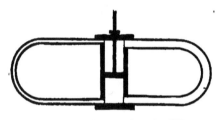

Fig. 41.—Diagram illustrating the difference of resistance between two water circuits.

and that this is varied according to the diameter and length of the wire. There are two kinds of true resistance that we may recognize. The first is specific resistance, by which is meant the resistance of a degree found in a wire of a given substance, say copper, of a given size and at a given temperature, as compared with that found in a wire of a different substance, say iron, of the same size and at the same temperature. The second kind of resistance we may call "comparative resistance," and it is computed between two wires of the same substance, but of different size. This follows the rules previously laid down.

Fig. 42.—Induction Coil of the type used in telephones.

In addition to true resistance, specific and comparative, there are several kinds of "false resistance," as the term is.

"False Resistance." — Under this head we have "impedance," or the resisting power of "self-induction" or induced magnetism, which modifies the efficiency of the current; also a constant interfering agency, known as "counter electromotive force." Both true and false resistance are occasionally computed together with the C E M F as integral parts of the particular resistance of a given circuit.

Electrical "Back Force" (C E M F).—As we have seen, the current given off from any electric source consists of what is known as electromotive force (E M F). The obstacle it meets on the wire, known as counter electromotive force (C E M F), behaves, in many respects, like a current moving in an opposite direction. It probably results from a combination of forces, and may be compared directly to the resisting agencies at work in the flow of water under head pressure, as is shown in

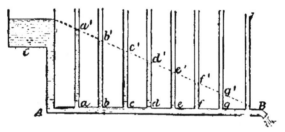

Fig. 43.—Diagram illustrating the effect of "back force" in water.

Fig. 43. Here *C* is a tank containing water, which flows out through the pipe, *AB*. If the vent at *B* is shut the water will stand at the same level in the tank and in the upright pipes, *a, b, c, d, e, f, g*. But if it is open, there immediately begins a a conflict between the natural tendency of water to seek its own level and the force of gravity, which causes the water to flow out of the vent. The result is that the water in the upright tubes stands at the levels *a', b', c', d',* etc. If the pipe, *AB*, is of smaller diameter than that of the uprights, *a, b, c, d,* etc., the resistance will allow of a greater effect in the counter water-motive force, which will keep the columns of water in them at a higher level than is shown in the figure. If, however, the diameter of *AB* is larger than that of the upright tubes, the columns of water in them will sink or disappear, showing that a decreased resistance in the vent-pipe is accompanied by a decrease in the "back-force." On precisely similar principles, the electromotive force of an electric current is so reduced over a long circuit that it is impracticable, in the present state of knowledge, to transmit *power* for running machinery and electric lighting to any great distance.

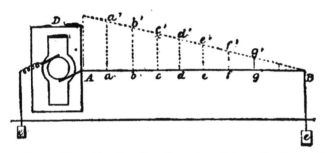

FIG. 44.—Diagram illustrating the effect of counter electromotive force; *D*, dynamo; *E E*, ground return. The principle involved is the same as in the preceding figure. If the live wire *A B* be tapped at any of the points *a, b, c, d*, etc., it will be found that the current regularly decreases in power as it moves further from the electrical source.

Electrical Power Transmission.—Thus, the vast power plant at Niagara Falls, with its chain of dynamos, each of 5,000 horse-power, is amply sufficient for the power and light needs of Buffalo and other near cities, but when, in 1896, a current was brought to New York City over the telegraph lines the power was no more than sufficient to run a tiny motor in the hall of the Electrical Exhibition. To have carried a force sufficient to be of commercial value, wires of immense diameter would have been required.

This "wasting away" of the E M F is one of the greatest difficulties in the way of "long-distance" telephony beyond a certain limit, although, as we shall see, the telephone receiver is an instrument of almost unimaginable sensitiveness, capable of being actuated by a current almost infinitesimal in power.

CHAPTER FOUR.

ELECTRICAL QUANTITIES.

Electrical Measurements.—Electrical measuremen.s are simple in theory, readily computed and of extreme exactitude. Starting with a few known quantities, we can get very exact figures for the strength of a given current, the degree of electromotive force any source can emit, and the resistance of any line of wire, as well as the amount of work it is capable of accomplishing in a given time, under any assumed conditions. Considering the need of exact instruments in such an industry as telephony, and the minute quantities of force and volume to be dealt with in them, it is indispensable that the practical telephonist have standards of measure that will permit him to deal with things and conditions almost unimaginably small.

An Exact System Necessary.—The first need in an exact system of measure is a standard of comparison with some known thing. Thus, when the French scientists invented the Metric system of measures, now so familiar, they sought to have as a basis a unit that could always be found from things known. In so doing they sought to find an "absolute" standard; that is to say, a standard of measure that would be the same the world over, and could be found at any time. All other units of measure have been variable and uncertain. The ancient Hebrews had a standard of measure called the "cubit," which was supposed to represent the length of a man's arm measured from the end of the elbow-joint to the tip of the middle finger. Their standard cubit was somewhat less than 22 inches, but the Roman cubit was only about $17\frac{4}{10}$ inches, indicating, possibly, a difference of over four inches in the standard stature of the Jews and the Romans,

The Metric System.—The decimal, or "Metric," system, adopted as the standard unit for measures of length the "meter," which, instead of being derived from the supposed length of a human limb, which is never the same in two people, or for two nations, is computed as one-ten-millionth of the total distance from the Equator to the North Pole of the earth. Now, according to the known laws of circles, we have ninety degrees as the total measure of one-quarter of any circumference; consequently, adopting as a standard the one-ten-millionth part of ninety degrees, we should be readily able to calculate this unit at any time.   But, because the earth is flattened at the poles, thus being an imperfect sphere, the geographical length of a degree decreases rapidly as we go from the Equator. However, the French astronomers who computed the matter, toward the end of the Eighteenth Century, seem to have estimated the average length of a degree of latitude at about 69.042 miles, which is something less than the length now estimated at the Equator: 69.16.

The Meter.—Having adopted as a standard the one-ten-millionth part of ninety such degrees, so as to have a decimal system of calculation, or one in the scale of tens, their unit was estimated as about .000621 miles; 1.09363 yards; 3.86090 feet; and 39.3709 inches.   This is the length now known as the "meter," or "metre," and from it all the measures of volume and weight have been derived.   The ease and convenience of having a scale of tens in measure greatly facilitates calculation, and enables one to find instantly the comparative values of any given quantity.

Decimal Units.—Thus a meter is sub-divided into ten parts, one hundred parts, and one thousand parts, giving the table, with Latin numeral prefixes, as follows:

> 10 millimeters make one centimeter,
> 10 centimeters make one decimeter,
> 10 decimeters make one meter.

These prefixes are the Latin words, *decem*, ten; *centum*, one hundred; and *mille*, one thousand. Measures larger than a meter are indicated by the Greek numerals: *dcka*, ten; *hekaton* (*hecto*), one hundred; *chilioi* (*kilo*), one thousand; and *myria*, ten thousand. Thus we have the table of quantities more than one meter, as follows:

  10 meters make one dekameter,
  10 dekameters make one hectometer,
  10 hectometers make one kilometer,
  10 kilometers make one myriameter.

Square, or surface measure, and cubic, or solid measure, are based on the same decimal scale, except that square measure naturally increases on a ratio of hundreds, as measuring two dimensions, and cubic measure, on a scale of thousands, as indicating three dimensions. Thus:

  100 square millimeters make one square centimeter,
  100 square centimeters make one square decimeter,
  100 square decimeters make one square meter, etc.

The cubic measure is as follows:

  1,000 cubic millimeters make one cubic centimeter,
  1,000 cubic centimeters make one cubic decimeter,
  1,000 cubic decimeters make one cubic meter, etc.

The Gram, or Measure of Weight.—Measures of weight and capacity are based on the cubic system. Thus, for weight we have the "gram," or "gramme," which is the weight of a cubic centimeter of distilled or perfectly purified water, weighed at the temperature when it has its maximum, or greatest density, which is at 4 degrees, centigrade, or 39.2 degrees, Fahrenheit, the latter being the common form of thermometer. Thus, as we would say, the water must be weighed when it is at a temperature of 7.2 degrees above the freezing-point, or "zero."

The Liter, or Measure of Capacity.—In the same fashion the unit of capacity is found by taking a box or can of the exact capacity of one cubic decimeter. This is what is called the "liter."

But the gram and the liter are divided and increased on a ratio of tens, instead of on one of thousands. Thus it is that

while a cubic centimer of water in the conditions just mentioned weighs one gram, a cubic decimeter of water would
weigh one kilogram, and a cubic meter, one hundred myria-
grams.   So, also, a cubic meter of any substance would occupy
a measure of one kiloliter, or one thousand liters.

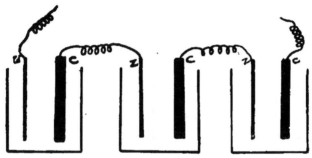

FIG. 45.—Electrical circuits, 1 : Diagram illustrating the connection of three galvanic cells in SERIES. The wire connecting the dissimilar poles of the several cells
passes through each in turn, making a negative terminal at one end of the battery of
cells, and a positive at the other.

Decimal Measure for Time.—At the time of the French
Revolution (1793–96), when the metric, or decimal, system of
weights and measures was invented, it was hoped that it would
be adopted as a universal method of calculation.   In this, how-

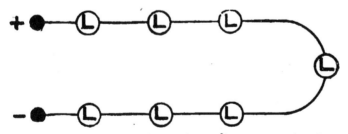

FIG. 46.—Electrical circuits, 2 : Diagram illustrating the connection of a number of
electric lights or other machines in a SERIES circuit. The poles at either terminal correspond to those of the battery, as in Fig. 45.

ever, expectations have not been realized, except in scientific
computations, where a common standard is positively necessary.
At the same time it was attempted to substitute a decimal
system for measuring time, instead of the duodecimal (twelve)

system now in use. But this change was so perplexing to the people that it was finally abandoned. Thus it is that while the new science of electricity uses the decimal scheme of measurements for all other quantities, it retains the old "twelfth-system" of seconds, minutes and hours.

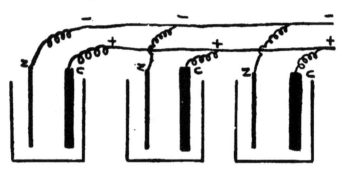

FIG 47:—Electrical circuits, 3 : Diagram illustrating three cells joined in. MULTIPLE or PARALLEL. The circuit is made by two wires, by which the like poles of all the cells are connected throughout.

C. G. S. Units in Electricity.

—All electrical measurements are based on the metric units of length (centimeter), of mass (gram), and of time (second). Since, as we have previously suggested, the quantities to be measured are so minute, we

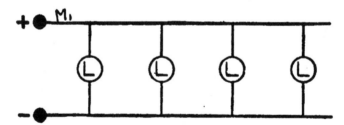

FIO, 48.—Electrical circuits, 4 : Diagram illustrating the method of connecting a number of electric lights, or other machines, in a MULTIPLE circuit. Like the cells in Fig. 47, they are strung on " bridges " between the two line wires, so that a complete circuit may be formed at every one.

must have the units as small as practicable. The centimeter, which is a little over one-third of an inch (.3937079 inch), not quite two-fifths of an inch, is capable of division into immensely small parts without the use of too many figures. The same is true of the grâm, which is employed in electrical measurements as the name for the cubic centimeter, without reference to weight, and is found most useful because it increases on a ratio of tens, instead of on one of thousands. The second, or time unit, is of use in computing the power of currents to do work, in connection with the length of lines and the total masses of wire in cross-section, as indicating the resistances of various kinds which the original supply of electromotive force must meet before the pressure can be used in any practical work.

Thus the units of electrical measures are spoken of as the "C. G. S." (centimeter-gram-second) units, as indicating that they are founded on the three simple units above mentioned, which give an absolute or universal standard.

Names of Electrical Units.—In addition to the need of units that may be expressed as readily and with as few figures as possible, the science of electricity further demands distinct names for its units in order that there may be no uncertainty as to what particular quantities are being measured. The names of electrical units of measure are based on the names of several great scientists and electricians of history, who thus have a lasting fame in the terminology of the science they labored to perfect. The electrical units are, accordingly, the "volt," named from Alessandro Volta, Professor in the University of Pavia, and discoverer of the electric current; the "ohm," from Prof. George Simon Ohm, of Berlin, a profound investigator in electrical matters; the "ampere," from Andre Marie Ampere, Professor of Mathematics in the Polytechnic School of Paris, and author of a celebrated work on dynamic electricity; the "coulomb," from Charles Auguste Coulomb, a brilliant investigator of electricity, and the author of a number of books on magnetism, the mariner's compass and general mechanics; the "farad," from Michael Faraday, the celebrated English physicist and Professor in the Royal Institution of London; the "watt," from James Watt, the inventor of the practical steam engine.

The Ohm, the Unit of Resistance.—The unit of electrical resistance is the "ohm," which, by the Electrical Congress at the Columbian Exposition, at Chicago, in 1893, was fixed as the equivalent of the resistance to the electric current of a column of liquid mercury, 106.3 centimeters (about 41.3 inches) long, and one square millimeter (.00155 square inch) cross-sectional area at the temperature of melting ice. Mercury is chosen for the test, because, as we have seen, its conducting power is low (1.6 in a scale of 100 for silver), and its specific resistance is comparatively high (99.74 in a scale with 1.521 for silver). It is, therefore, about one on the scale when measured for conduction; and about one hundred on the same scale when measured for resistance. The standard specifies the temperature of melting ice, since, as is known, a column of mercury is shortened or lengthened with the fall or rise of temperature, and on this fact the thermometer is constructed.

Electrical Pressure.—The fact that there is such a thing in the electrical conductor as resistance, which may be measured and must be met by a force which is able to do work, establishes the further fact that there is such a thing as *pressure.* The steam engine works on this principle: if there is a sufficient pressure of steam generated in the boiler the machinery may be moved; otherwise, not. Similarly, if we have a sufficient "head" of water, we may move the wheel of a mill. In both cases there is a power or force in the steam or the water that can overcome a resistance in matter, which is known as "inertia," or the tendency to remain motionless unless compelled to move. Now, the electrical element that can set up an active condition in a circuit of metal, possessing the electrical inertia, called resistance, is the electromotive force (E M F) that we have previously mentioned. It is for electricity what the steam pressure is for the engine, and what the head of water is to the mill-wheel.

The Volt, the Unit of Pressure.—The unit of electrical pressure is called the "volt," and it is about equal to the average strength of an ordinary galvanic cell. Thus, the usual type of Daniell (or "blue-stone") cell used in telegraphy has a capacity of about one (correctly, 1 08) volt. The Leclanche cell, which is the kind most often used in telephony, has a capacity of one and one-half (1.50) volts. One volt of E M F

passed through a circuit having a resistance of one ohm yields one unit of working electrical energy. Such a unit of resistance may be found in one foot of No. 40, A. W. G. (American Wire Gauge) wire, which has a diameter of three-one-thousandths of an inch (003145 inch, or 3.145 mils) The same unit of resistance is to be found in about two miles of ordinary copper wire used in electric trolley lines. In both cases we have just about the equivalent of the column of mercury just mentioned if we have also a temperature of 45 degrees, Fahrenheit.

F I G. 49.—C o m m o n "crow-foot," blue-stone Daniell cell. Negative electrode, a claw-shaped piece of zinc; positive electrode copper plates, among which is disposed blue vitriol (copper sulphate).

The Ampere, the Unit of Current.—The unit of current intensity given forth after one volt of electromotive force has passed through a resistance of one ohm, is called an "ampere." Its strength has been found to be equal to that of a current which can deposit .00033 grams (that is, thirty-three-one-hundred-thousandths of a cubic centimeter) of pure copper in one second of time by the process of electro-plating. Because of this known strength of the ampere in terms of the electroplating process, we have such electrical current meters as the Edison "ammeter" (ampere-meter), the construction of which is shown in Fig. 50. As may be seen, the current moves along, as is indicated by the arrows, through the two binding-posts and across the bent or twisted section. A portion of the current is "shunted," or switched off from the main line, and flows through a wire of known resistance into the cell, where it acts on the principle of an electro-plating machine, carrying small particles of the incoming electrode (the "anode") to the outgoing (the "kathode"), as the current passes through the liquid solution in which they are immersed. As the shunted current is in direct and known proportion to the main current, the amount of metal deposited in a given time is an indication of the total amount flowing through the line. This may be determined by simply weighing the negative electrode.

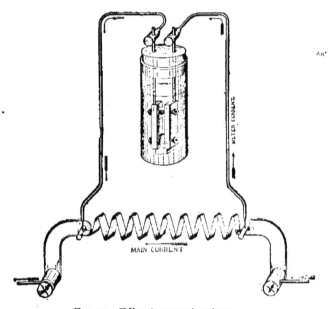

FIG. 50.—Edison's current meter.

Ohm's Law.—Because, as we have seen, the resistance of any wire of given length and thickness is known, because the current may be measured by efficient instruments, and because there is a known ratio of voltage in most common battery cells of given size, we have in every circuit at least two known quantities, which will enable us to determine the other. The relation of all these quantities to a given standard—since one volt, after passing through one ohm, will yield a current of one ampere—establishes a certain proportion among them. This may be expressed in the formula known as Ohm's Law, which is as follows:

I.—The current is in direct proportion to the electromotive force (E M F), and in inverse proportion to the resistance.

This is evident, because it is by the resistance of the conductor that the initial electrical pressure of the source or battery is *modified*, or divided; one percentage being, supposedly, "dissipated," or "absorbed," as is steam power or water pressure, and the other given forth as working current.

With an understanding of these facts, we may see that the rule given above means, in simple language, that: The greater the electromotive force, or the smaller the resistance, the greater the current; and the smaller the electromotive force, or the greater the resistance, the smaller the current. Moreover, because, as we have seen, there is a definite proportion between the three units in any circuit, the ratio of the three is precise and not general. Thus, we may derive the rule:

II.—The current (amperage) is equal to the electromotive force (voltage) divided by the resistance (ohmage.)

This rule may also be expressed as follows:

$$\text{Current} = \frac{\text{E M F}}{\text{Resistance}} \text{ or Ampere} = \frac{\text{Volt}}{\text{Ohm.}}$$

Therefore, knowing the voltage of the source and the ohmage of the circuit, the current strength may be determined by dividing the former by the latter.

III.—The resistance varys directly with the electromotive force, and inversely with the current.

This is a law of proportions precisely similar to the first rule given, and, like it, may be reduced to a more exact form, as follows:

IV.—The resistance (ohmage) is equal to the electromotive force (voltage) divided by the current (amperage.)

This is equivalent to saying: The ohm is the quotient of the volt divided by the ampere, or:

$$\text{Resistance} = \frac{\text{E M F}}{\text{Current Strength.}}$$

V.—The electromotive force (E M F) varies directly with the current and with the resistance.

This rule may be readily understood by reference to the first and third of the series, wherein it is stated that both the current and the resistance vary directly, or just as does the electromotive force. Consequently, the formula is readily understood:

VI.—The electromotive force is equal to the current multiplied by the resistance.

In tabulated form, this statement may be expressed as follows:

E M F = Current × Resistance, or
Volt = Ampere × Ohm.

Ohm's Law Applied.—As may be readily understood by careful study, all these rules are merely so many variations of one statement of proportions between the three factors of any active electrical circuit. They are, in fact, corollaries of the first rule, which is a sufficiently complete statement of Ohm's law for general purposes.

A few illustrations will suffice to make the applications of the law perfectly clear. If, for example, we have two active electrical circuits with precisely the same current strength, we know that one of three things must be true; either,

1. They are the same in all respects, as regards voltage, ohmage and amperage; or,

2. The original voltage is greater, involving that the resistance, both internal and external, must be greater also; or,

3. The original voltage is smaller, involving that the resistance, internal and external, must be smaller also.

If, in any circuit, the original pressure is 8 volts and the resistance 4 ohms, it follows that the current strength is 2 amperes. Now taking these figures as the basis of equations, instead of the unit-names, as above, we have :

(current) $2 = \dfrac{8}{4}$

(ohmage) $4 = \dfrac{8}{2}$

(voltage) $8 = 4 \times 2.$

Water Currents and Electrical Currents.—In many points the behavior of water currents is comparable to that of electrical currents. The three elements—pressure, resistance, efficient force per second—are also present. Thus, if we have a body of water confined in a convenient tank, or other receptacle, so that a constant stream may escape through an orifice one inch square, under the pressure of six inches—this is measured from the bottom of the tank to the surface of the water—we have the equivalent of what is known as the "miner's inch," an exact analogy to the ampere.

Taking the pressure of the column of water in the tank, which increases directly with its height and the dimensions of the orifice through which it escapes, we find that the miner's inch is, first place, a measure of rate or velocity, a time measure, giving inch-seconds. Thus the water flows at the rate of so many miner's inches, just as electricity flows at the rate of so many amperes, both in direct proportion to the initial pressure. In the water current the size of the orifice of escape measures the resistance; in the electrical current it is the size, or diameter, of the wire, as we have seen, provided it be compared with another wire of the same metal. Thus it is that in neither case do we have a reference to time. An "ampere per second" is a simple repetition, and means no more than an "ampere." Thus, in speaking of a current of, say, ten amperes, we do not refer to the current passing in ten seconds, but to that passing in one second.

The Coulomb, the Unit of Electrical Quantity.—This brings us to another unit of electrical measurement, the measure of electrical quantity. This is the "coulomb," and it is equivalent to the efficient quantity of electrical energy passed by a current of one ampere intensity in one second. It is, in fact, the "ampere-second." Experimentally it has been found to equal the decomposition of .0936 milligrams of water, by electrolysis, or the deposition of 1.11815 milligrams of silver by electro-plating. In both cases, of course, the actuating current is equivalent to one ampere, and the initial pressure, to one volt.

The Watt, the Unit of Electrical Activity. —A fourth unit is the "watt," which is the practical unit of electrical activity. It is the rate of energy or of work represented by a current of one ampere urged by one volt of

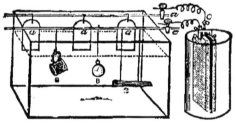

FIG. 51.—Diagram illustrating the method of electro-plating. *a* is the anode joined to the positive side of the cell; *c*, the kathode joined to the negative side; *B, B, B*, articles intended for plating by *ions*, or particles, carried by the current from *a, a, a*.

electromotive force. It is, in fact, the volt-ampere, or the product of the pressure and the current. Applying the principles of Ohm's law, given above, we find that it is, further, the product of resistance multiplied by the square of the current, and the quotient of the square of the voltage divided by the resistance. If, then, we have a circuit with a voltage of 8, an ohmage of 4, and a current of 2, we may compute the wattage as follows:

(voltage) 8 × (current) 2 = (wattage) 16.

(resistance) 4 × (square of current) 4 = wattage 16.

(square of the voltage) 64 ÷ (resistance) 4 = wattage 16.

The Equivalent of the Watt.—The wattage of a given active circuit is equivalent to the product of the voltage and the amperage, because, as is evident on reflection, the efficiency of any current for the accomplishment of work depends upon the pressure that impels it. For example, the energy furnished per second by a current of 10 amperes supplied at a pressure of 2,000 volts is exactly the same as that furnished per second by a current of 400 amperes supplied at a pressure of 50 volts.   In each case the product is 20,000 watts.

Watts and Horse Power.—Approximately 746 watts is equal to one electrical horse power.   A watt-hour is the expression for a power of one watt exerted by a current during the space of one hour.   The watt-minute represents the same power exerted during one minute.   The watt-second is equivalent to the product of the volts and the coulombs of a given circuit, and is also known as the volt-coulomb.

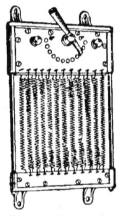

Fig. 52.—Common type of resistance frame. By moving the lever on the row of studs, one or more of the coils of iron, German silver, or other high-resistance metal, may be thrown into circuit ; thus varying the line resistance at will.

While the measurements of electrical quantity and activity have no bearing on telephony, except in the questions connected with the use of charging dynamos and central energy units now coming into use more and more in up-to-date exchanges, it is essential that we include them in this brief survey of the system of electrical measurements, in order to give a good foundation for future and more extended study of practical telephone matters.

Electrical Condensers.—One consideration of importance, both in telephony and in other branches of electrical industry, is that of condensers.   The condenser, or accumulator, is an instrument capable of holding a charge of electricity, and of giving it forth in the form of a shock or momentary current.   More correctly, it is a device capable of being so affected by a primary electrical source that its internal potential is raised proportionately, and will electrically affect any body of lower potential.

The simplest form of condenser is the Leyden jar, a glass jar, coated within and without with tin foil.   A more practical variation of the same principle is the ordinary telegraphic condenser, which consists of a number of sheets of tin foil, each two separated by a sheet of paraffined paper or mica. The tin foil sheets are connected to the poles of the primary battery, the first, third, fifth, and all odd numbers being connected together by a wire leading to one pole, and the second, fourth, sixth, and all even numbers, by a wire leading to the other pole. This arrangement is shown diagramatically in Fig. 53.

FIG. 53.—Diagram illustrating the principle of the common electrical condensor.

Construction of Practical Storage Batteries.—The form of device used largely in electric lighting, and also in central energy telephone exchanges, is usually described as the "storage battery," and depends for operation on chemical action. One of the earlier forms of this device was the Planté secondary battery. It consisted of pairs of lead plates, arranged in the order described for the ordinary condenser. They were placed close together, but not touching, in a solution of sulphuric acid. The battery was then "formed" by the action of an electrolyzing current in one direction. By this process the surface of one plate is largely converted into lead binoxide. By frequent repetitions of the process of alternately charging and discharging this secondary battery, both plates are considerably attacked, one of them, however, being maintained in a state of oxidation. Later forms of the secondary cell have the lead plates or "grids" perforated, as shown in Fig. 54, and these perforations are filled with some substance suitable to enable the grid to act at once as either positive or negative electrode, as the case may be, thus giving a good depth of matei.a for the charging current to act on, and avoiding the tedious process of "forming," in order to prepare the cell for practical use.

The Chloride Storage Cell.—Fig. 55 gives a good general idea of a type of storage cell much used in central energy telephone exchanges. It is charged in the same manner as the Planté cell, "after formation" but is much superior in construc-

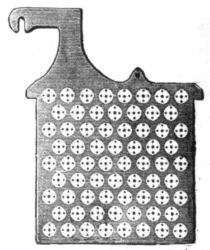

FIG. 54.—Single "grid" of a common type of storage battery.

tion and efficiency. As will be seen in the figure, the alternate grids have square or round perforations. The former are the negative and the latter the positive electrodes. The positive grids are composed of an alloy of lead and antimony cast into shape with the required number of round perforations. Each of these holes is filled with a button made by rolling a crimped lead ribbon in the form of a spiral and of a size to fit it tightly. The plates are then treated electro-chemically in order to produce the proper lead oxide. The negative grids are made by casting the proper shape, under heavy pressure, around a number of square blocks of fused chloride of zinc and chloride of lead. After the grid is thus completed the zinc is chemically removed, leaving the contents of each perforation pure spongy lead. To assemble the cell, the grids, positive and negative, are hung alternately in the tank or box, being held apart by sheets or washers of hard rubber. A solution of sulphuric acid, about five parts water to one part acid, is then poured into the tank, and acts as the electrolyte, after the full charge has been given from a dynamo.

Action of a Storage Cell.—The action of the cell consists, first, in the oxidation of the spongy lead of the negative plates. The hydrogen set free in this process is transferred electrolyti-

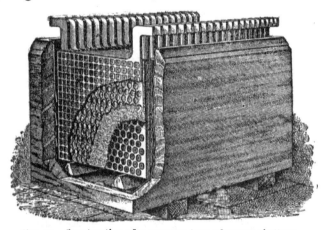

FIG. 55.—Construction of a common type of storage battery.

cally to the positive plate, where it reduces the lead dioxide (Pb O'') to lead protoxide or monoxide (Pb O'). The oxides are then converted into sulphates by the action of the electrolytic solution. In the process of charging the conditions are reversed under the action of the electrical current from the dynamo, which has the effect of reducing the lead sulphate on one plate to metallic lead, and that on the other to lead binoxide, or dioxide. As soon as this process is completed the storage battery is said to be "charged," and is once more in condition to give forth a current. Although the storage, or secondary, battery differs radically from the electrical condenser, despite the popular identification of the two, it is proper to regard it as a form of accumulator—since a certain amount of current is required to "form," or charge it, as we have seen—and to measure its power in terms of electrical capacity and quantity.

The Farad, the Unit of Electrical Capacity.—The theoretical unit of electrical capacity, as applied to condensers and accumulators, is the farad, which represents the capacity of a conductor which can retain one coulomb of electrical energy at a potential of one volt. The quantity of electricity charged upon a conducting surface, however, raises its potential, so that a condenser of one farad capacity can hold two coulombs at two volts, or three coulombs at three volts.

Every Line a Condenser.—In practical computations the microfarad, or one-millionth of a farad, is used. The faradical capacity of the earth is estimated as $\frac{636}{1,000,000}$ of a farad. As has been well said: "Every line is practically a condenser, the surface of the wire being one plate, the earth the other, and the air or covering of the wire the insulating layer between the two. Before a current can reach its full strength at the distant end of the line this condenser must be fully charged; and when reversed currents are used it must be discharged and charged again at each reversal. This explains why a high capacity is so detrimental to telephone working, as the rapid changes and reversals of the telephonic current are deadened and flattened out if they have to charge much capacity before reaching the receiving instrument."

The Use of Electrical Units.—Electrical units, while computed on the basis of the metric, or decimal system, are usually expressed in fractional forms, with the prefix, micro (Greek, *micros*, small), indicating the one-millionth part of a given unit, as micro-farad; or, similarly, with the prefix mega (Greek, *megas*, great), as indicating one million times that unit, as megohm, etc. The practical units used in computation are usually fractions of those given. But, as the exact description of the system belongs to a more extended treatment of the subject than is here necessary, it will be omitted.

CHAPTER FIVE.

HISTORY OF THE SPEAKING TELEPHONE.

When we are told that the steam engine was invented by James Watt, or the steamboat by Robert Fulton, or the steam locomotive by George Stephenson, or the telegraph by S. F. B. Morse, or the telephone by Alexander Graham Bell, we are very apt to suppose that each of them originated both the idea and the contrivances embodied in their several inventions, and that no one ever thought of any such things before. As a matter of fact, however, these and all the other great inventions of the Nineteenth Century are merely the perfected product of many minds working during many years, although the man who finally completed the labor gets all the credit.

Almost from the time when electricity began to be of practical worth in the world there were some people who were at work on the idea of perfecting contrivances for the electrical transmission of speech and other sounds. And in this department of endeavor, as in others, we find that the final and perfect instrument is the simplest.

Reis's Telephone.—One of the earliest experimenters to realize successful results was Philip Reis, Instructor in Natural Sciences at Prof. Garnier's Institute, a select school for boys, at Friedrichsdorf, near Homburg, Germany. His principle has frequently been described as "making and breaking" an electric circuit, in somewhat the same fashion as the electric telegraph transmits messages; but Prof. Sylvanus P. Thompson, in a paper on Reis's invention, read before the Naturalists' Society of Bristol, England, in 1883, declares that it was rather the "employment of a loose or imperfect contact between two parts of a conducting circuit," so that the pressure and resistance might be varied by differing stress.

67

On the occasion of reading this paper he exhibited a model
of Reis's original transmitter, which was shaped like the human
ear, and was designed to work on the same lines.   Fig. 56
shows a section of this instrument.   Here we have an opening,
A, corresponding to the meatus of the ear, and closed at the
"inner" end with a membrane, *B*, corresponding to the tympanum, or drum.   Immediately against this membrane rested

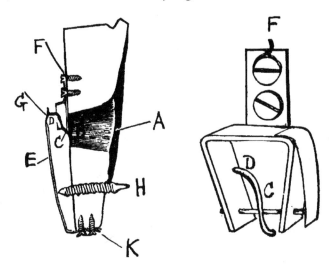

FIG. 56.—Reis's first transmitter.   FIG. 57.—Detail of lever, Reis's
                                         transmitter.

the lower end of a curved lever of platinum wire, *C*, which
represented the "hammer bone."   It was attached to the
membrane with a drop of sealing wax, and was left free to
move, as the membrane vibrated, by being soldered to a short
cross-wire, serving as an axis, as shown in Fig. 57.   The upper
end of the lever, *D*, rested against the end of a vertical spring,
E, which bore on its extremity a slender and resilient strip of
platinum foil, *G*.   The contact of the spring could be varied, as
occasion demanded, by the screw, *H*.   The electrical connections were made by the wires, *F* and *K*, the former bearing a
current to the lower end of the lever, through its connection
at the center of the diaphragm, and the latter transferring to

the line the sound-impulses caused by varying the pressure of the lever on the strip of platinum, *G*, as it moved with greater or less force in obedience to the vibrations of the diaphragm, according to the motions of sound waves.

A later form of the Reis transmitter was a box having two openings, one at the side for the mouthpiece, and the other at the top, closed by a diaphragm made from the smaller intestine of a pig. At the center of this membrane was cemented a strip of platinum in loose contact—some say not touching—with the point of a platinum wire held in position above it by a leaf spring, as in the former instrument.

Reis's Receiver.—The receiver of the Reis system was a steel knitting-needle wrapped around with a coil of insulated wire. It was, in fact, a form of electro-magnet, and was mounted on a violin or resonant box, which served as a sounding-board, and to this was later added a cover against which the ear was pressed to receive the sounds transferred along the wire from the transmitter. By the Reis system musical sounds could be transmitted with all the variations of pitch and loudness, although without timbre—probably resembling somewhat the sound of a xylophone, or wooden piano—and less perfectly, also, the sounds of the human voice; the consonants being readily represented, but the vowels less distinctly, if at all.

Prof. Reis anticipated many of the later features of the practical telephone, particularly the transmission of sound by the variation of pressure in an electrical circuit, and also made his receiver on the plan at first adopted by Prof. Bell—an electro-magnet and a resonant diaphragm. His instruments attracted wide attention in that day, but no one thought of applying them to any field of commercial usefulness, most probably because the transmission of speech was too imperfect, even if the social and business conditions had created a demand for such a contrivance. During the next sixteen years little was heard of the subject of transmitting speech by electricity, and when, in 1876, Prof. Bell exhibited his imperfect instru-

ments at the Centennial Exposition in Philadelphia, there was
another sensation among scientists.

House's Telephone.—Meantime, however, the musical telephone was much improved, and several independent experimenters actually made some progress toward a practical speaking apparatus. Among the latter was Royal E. House, of
Binghamton, N. Y., who in 1868 invented and patented a
device which he called the "electro-phonetic telegraph." Its
general construction included a sounding-chamber, open at the
front end and closed at the rear by a diaphragm of varnished

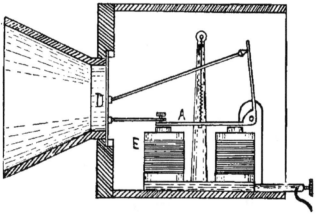

FIG. 58.—Diagram of House's "Electro-phonetic Telegraph." D is the Pine Wood
Diaphragm ; E, The Electro-magnet ; A, The Armature.

pine wood, to the center of which was attached a bar, with the
object of imparting the vibrations of the diaphragm to the
upright piece attached to the horizontal armature of an electro-
magnet. Two instruments of this pattern at either extremity
of a circuit, which includes a battery cell, can act alternately as
transmitter and receiver, and actually convey spoken and musical sounds. Mr. House, however, seems to have had no
conception of the value of his invention; so he rested content
with local celebrity, leaving the fame and rewards coming from
the successful telephone to be gathered by another man eight
years later.

Gray's Telephone.—Prof. Elisha Gray was also conducting experiments in telephony, and, strangely enough, entered an application for a transmitter very like Bell's in general theory on the very day a patent was issued to the latter.

Drawbaugh's Claims.—In 1884, in a suit for infringement of patent by the Bell Telephone Co. against a concern known as the People's Telephone Co., brought to light the claims of one Daniel Drawbaugh, a machinist, of Eberly's Mills, Penn., to having invented a complete and successful telephone apparatus at least ten years before the date of Bell's patent. Drawbaugh sought to establish his claims by the testimony of 145 witnesses, who swore that they had seen the machines in operation in his shop or heard them spoken of by others between the years 1867 and 1876. He also exhibited a series of models, the earliest of which he claimed was made in 1866, and the latest about ten years after. His transmitters were constructed on the principle of the carbon granule instruments subsequently adopted by Hunnings and others, and his receivers were of the magnet type, now in use on all lines. His claims, if true, certainly present a strange case of almost perfect coincidence. But it was this fact, as much as any other, that moved the decision of the United States courts against him.

In one of his experimental model receivers, dated about 1866, the diaphragm is connected to the armature of an electro-magnet by a taut string, a device adopted by Preece as late as 1879. A still later one (1867-68) has the armature of the magnet on the diaphragm, as is the case with Bell's instrument, made early in 1876. About 1870 he claimed to have devised a double-pole permanent magnet telephone, identical with that later patented by Bell, and to have further improved it by the use of spiral magnets, with a single-pole contact, as in some of the more modern forms of receiver. His case certainly deserves a passing mention.

Bell's Early Instruments.—Prof. Bell seems to have made a large number of experiments with almost as many different

combinations before he finally hit upon the wonderfully simple
instrument patented in 1876.   One of the earliest of his
experimental instruments is shown in Fig. 59.   It consists of an
elongated type of permanent horseshoe magnet, on each pole of
which is attached a harp of rods, like those used in the familiar
mechanical music boxes, each rod in the comb-like contrivance
being cut to a length suitable to giving forth its own note.
Now, when one sang near this instrument, he effected a vibration in the key corresponding to the note sounded by his voice,
thus producing a "wave" of a size, or "amplitude," proportioned to that uttered.   This would produce an agitation in the

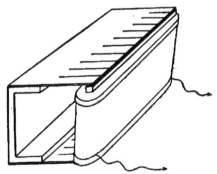

FIG. 59.—Bell's First Experimental Instrument : An Elongated Permanent V-shaped
Magnet, Bearing a Harp of Reeds, which Vibrate in the Field of an Electro-magnet.

field of the electro-magnet, in which a magnetically induced
current was at once set up.   The current, or series of currents,
thus induced, traveled on the line wire to the coils of the other
electro-magnet, which, in turn, attracted the corresponding key
of the harp, with the same strength and shape of wave as that
which left the "transmitter," thus reproducing the sound by
reversing the process.   The theory of operation has been briefly
expressed as follows: "The strength of the induced currents is
determined by the amplitude of the disturbing vibration, and
the amplitude of the vibration at the receiving end depends
upon the strength of the induced currents."   It would seem,
from the description of this instrument, that it presented the
same difficulty as Reis's—imperfect transmission of the vowel

sounds. It was also impractical from the vast expense of manufacture. It is of interest, however, as representing the earliest form of the theory adopted by Bell.

In his next experimental instrument Prof. Bell discarded the permanent magnet and induced currents, and adopted as his transmitter the type of instrument shown in Fig. 60—an electro-magnet mounted on a post by an adjusting screw, so that it may be approached to the vibrating diaphragm. This diaphragm is made of goldbeater's skin, and bears at its center nearest the magnet a small piece of soft iron to act as an armature. This diaphragm, when vibrated by a sound, produced a variation of the current in the coils of the magnet by disturbing

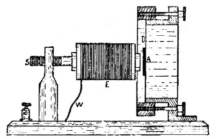

FIG. 60.—Bell's Centennial (1876) Transmitter.

the field of force. This process enabled the current to transmit variations in the strength and amplitude of its vibratious to the receiver which could retransform them into waves of sound. The receiver was another electro-magnet, in this instance consisting of a wire coiled about an iron tube. About this was an iron sheath, which was closed at its upper end by a thin plate of sheet iron lightly laid upon it, so that when no current was passing it would be just above the pole of the magnet. The vibrating current was able so to act on this diaphragm that the sounds varying the current in the receiver would be produced by similar currents, alternately attracting and releasing it. Such were the instruments exhibited at the Centennial Exposition in 1876.

Bell's First Permanent Magnet Receiver.—Later in the same year Bell discarded the electro-magnet and battery, and returned to his original idea of a permanent magnet and induced currents resulting from disturbances in its field by the vibration of a metal diaphragm. What such disturbances of the magnetic field of force are capable of accomplishing we have already seen. Bell's magnet telephone is, in fact, a miniature dynamo, and the sound-vibrations of its diaphragm transmitted along a line wire by the current they generate affect the receiver, which is precisely similar in construction, as a commercial current affects an electric motor. The receiver is thus, for a time, a motor, and when itself employed as a transmitter it becomes a dynamo.

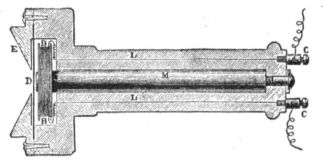

Fig. 61.—Bell single pole magnet telephone enclosed in wooden case.

Bell's final experimental instrument consisted of a powerful permanent horseshoe magnet, at each pole of which was attached a bobbin carrying a coil of fine insulated wire. Directly in front of these coils is the diaphragm—a thin iron disc supported in a wooden frame in such a manner that it can vibrate freely in obedience to the sounds uttered against its face through the mouthpiece. The relative distance of the diaphragm and magnet may be varied by clamps.

He also used a bar magnet, and opposed but one pole to the diaphragm. In this instrument the final form of the Bell telephone was reached. It only remained to make a suitable case for it, as shown in Fig. 61, and this need was supplied by the ingenuity of Prof. Pierce, of Brown University, Providence, R. I. In this figure, *M* is the bar magnet, *B* the wire-wound bobbin, L and L the line wires carrying the speech-bearing current through the binding-posts, C and C; D is the iron diaphragm, and E the mouthpiece of the case. The present type of magnet telephone differs in no essential particular from this instrument, except in the fact that the mouthpiece is now screwed to the head of the case by engaging a thread on the periphery, as will be noted in some of the figures of modern instruments given in the next chapter.

Superiority of the Permanent Magnet.—The permanent magnet is superior to the electro-magnet in a telephone instru·ment; first, because it dispenses with the battery cell in the circuit, and second, because in the use of an electro-magnet the magnetic force and the stress due to the periodic vibrations —contacts and withdrawals—of the diaphragm vary in proportion, so as to compel the diaphragm to execute double vibrations in the production of any one sound. The double vibrations,

Fig. 62.—Usual form of hard rubber receiver case.

while not observable, would have the distinct effect of rendering the reproduction of speech in the receiver less distinct. This was the case in the Bell instruments based on electro-magnets.

Simplicity of the Magnet Telephone.—In considering the extreme simplicity of the Bell magnet telephone one cannot but be surprised that it was not thought of sooner, particularly when we learn that the principle on which it works was recognized years before Bell ever dreamed of the electrical transmission of speech. Thus, as early as 1837 a Dr. Page, of Salem, Mass., first demonstrated the principle, that if a bar of iron be wound with wire, through which a current is passed, a distinct series of sounds will be emitted by it every time it is magnetized or demagnetized. This is the basis of all receivers, from Reis to Bell. Page and his contemporaries attributed the sounds thus made to "molecular disturbances, whereby the bar was lengthened."

Somewhat later a Mr. Farrar, residing in New Hampshire, devised a receiving instrument, consisting of an electro-magnet whose armature vibrated in the center of an elastic disc or diaphragm, and thus received musical sounds—he never thought of speech—from a distant transmitter, constructed on the plan of Bell's original experimental instrument, the several reeds or rods being set in vibration by a keyboard, at the same time opening or closing a circuit. Prof. Dolbear, of Tufts College, Mass., has described this instrument in a lecture in 1882, and compared the process of opening and closing the circuits to Helmholtz's tuning-forks, later devised to accomplish the same result by their vibrations, or to some of the harmonic devices now employed in quadruplex telegraphy—the method of sending four messages over the same wire at the same time. The principle of harmonic vibrations, whereby, as we have previously seen, one string or bell of a certain note may call forth vibrations from another of the same note by merely sounding in its vicinity, is the one on which the modern system of "wireless telegraphy" is based.

Prof. Bourseul's Prophecy.—In 1854 Charles Bourseul, a French scientist, published a paper on the possibility of transmitting speech by electricity, in which he says:

"Suppose a man speaks near a movable disc, sufficiently pliable to lose none of the vibrations of the voice, and that this disc alternately makes and breaks a current from a battery; you may have at a distance another disc, which will at the same time execute the same variations. * * * * It is certain that in a more or less distant future speech will be transmitted by electricity. I have made experiments in this direction; they are delicate, and demand time and patience, but the approximations obtained promise a favorable result." This favorable result was not achieved by Bourseul, however, but was delayed for twenty-two years, until Alexander Graham Bell perfected the telephone with the simplest possible electrical contrivance which could embody the principles laid down by the Frenchman.

Sensitiveness of the Magnet Telephone.—Not only is the Bell instrument a model of simplicity, but its degree of delicacy is simply amazing. This fact has been shown by a number of experiments described by Preece, the noted English telephonist.

Siemens' Experiment.—The first mentioned was by Dr. Werner Siemens, of Berlin. He took a Bell telephone, the pole of whose magnet was wound by 800 turns, or convolutions, of insulated copper wire, one millimeter (.03937 inch) in diameter, and of 110 ohms resistance, and placed it in the circuit of one Daniell cell (1.08 volt). By means of a mechanical commutator, placed in the circuit, the current from the cell was reversed 200 times per second, thus giving a series of impulses which produced a loud crackling noise in the telephone instrument. By use of Ohm's Law, previously explained, we find that a current of 1.08 volt passing through a resistance of 110 ohms is equal to .00981 amperes, or, approximately, .01 ($\frac{1}{100}$) ampere. In other words, it is such a current as would deposit about .0000033 grams (about thirty-three-ten-millionths of a cubic centimeter, or about two-ten-millionths of a cubic inch) of pure copper in one second, by the process of electro-plating.

Such a current has an intensity of almost unimaginable minuteness, but Dr. Siemens further modified it by inserting a resistance of 50,000,000 ohms, and placing the primary winding of an induction coil in the battery circuit, the secondary winding being attached to the telephone. This added resistance must certainly have reduced the current strength to a point about equal to .000,000,2 (two-ten-millionths) ampere in intensity. But the induced current was capable of maintaining a loud noise in the instrument, which still continued audible even when the secondary coil was pushed out clear to the end of the primary, thus reducing the inductive action of the primary coil to the minimum, and placing it beyond the reach of measurement.

Preece's Experiment.—Mr. Preece himself determined by a somewhat similar line of experiments that the Bell instrument will reproduce a sound actuated by a current equal to .000,000,000,000,6 (or six-ten-trillionths) of an ampere in intensity, or such a current as could deposit about .000,000,000,000,03 (or three-one-hundred-trillionths) of a cubic inch of pure copper in each second of time. Similarly, by using an electrical condenser of one microfarad capacity, which was discharged

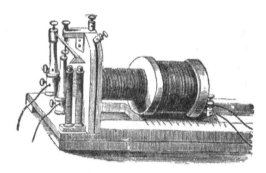

Fig. 63.—Induction coil, or inductorium, arranged with device for withdrawing the secondary coil from the primary. This instrument is provided with a mechanical rheotome, or circuit breaker, which reverses the current from the battery, as mentioned in Siemens's experiment.

through the circuit of a Bell telephone 160 times per second, a French electrician, Pellatt, calculated that, with a voltage of .0005, representing the difference in potential, or electrostatic capacity, between the two terminals, an audible sound could still be maintained in the receiver. Such a voltage of electromotive force would be of sufficient strength, as may be found by calculation, to produce one gramme-degree of heat in ten thousand years. That is to say, it would take ten thousand years for such a force to raise one gram, or cubic centimeter, of distilled water from zero, Centigrade, to one degree above, provided that the water retained every atom of heat imparted to it in the meantime.

Still another interesting experiment is recorded in Houston and Kennelly's work, "The Electric Telephone." These authors lay down the rule that the sensitiveness of the Bell instrument to feeble alternating currents has been estimated as best determined with a current of 640 alternations per second. This number corresponds, approximately, to the frequency of vibrations of the E note in the second octave below "middle C" in the piano scale. At this frequency, an alternating current strength of .000,000,044 (or forty-four one-billionths) ampere has been found able to produce distinctly audible sounds. Whether this is considered the limit or not, it is calculated that the force required to maintain this current strength through the resistance of the usual telephone receiver, which is about 75 ohms, is so exceedingly small that the work done in.raising a weight of thirteen ounces through a vertical distance of one foot, would suffice to keep an audible sound in the instrument for 240,000 years.

Such figures prove, either that the Bell telephone is the "most beautiful illustration of the equilibrium and unity of natural forces in existence," or else, that the standards of electrical measurements are on an entirely wrong basis. They are, however, of interest as illustrating the curious facts regarding the perfection of this simple contrivance, although of little account in the questions connected with practical telephony.

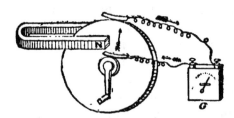

Fig. 64.—Faraday's electrical machine, the earliest dynamo constructed.

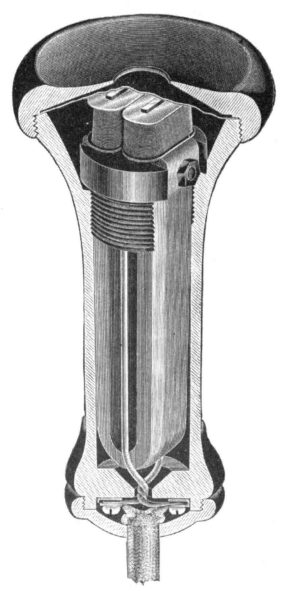

FIG. 65.—Bi-polar receiver of the Western Telephone Construction Co. The adjust-
·ent between the magnet coils and the diaphragm is made by the screw thread on the
ass block below the coil bobbins. The method of attaching the cord has the advantage
positively preventing contacts of the wires, and short circuiting.

CHAPTER SIX.

LATER MODIFICATIONS OF THE MAGNET TELEPHONE.

In the experiments just recorded, for testing the marvelous sensitiveness of the magnet telephone, we have found that the best results were obtained with alternating currents—those that are made to flow first one way and then the other—and that the telephone instrument was used as an electro-magnetic motor, rather than as a dynamo. The difference, as we have seen, depends on whether the instrument is used as a transmitter or as a receiver, since the motor differs from the dynamo only in the fact that it is made to move by the currents which the latter produces. Agitation of the magnetic field of force by the steady shifting of the armature produces a current, which, in turn, may agitate another magnetic field and thus cause the armature to move in the same way.

Conditions for Transmitting Speech.—In order to improve the practical telephone to the highest point of efficiency two things are necessary, as we may readily understand: First, that the currents bearing articulate sounds should be, as far as possible, of the alternating type—such has been found most effective for distinct transmission; and second, that the magnetic field in the receiving instrument should be as strong as possible, with every point of its available space utilized. The first of these needs has been ably met in the numerous forms of carbon microphone transmitters, mounted with a voltaic circuit and induction coil. The second has been almost as well supplied by the numerous improvements in magnetic receivers, although in general practice the original pattern instrument, of either one or two-pole contact, is still in common use. Magnet telephones are most often used only as receivers, on account of their extreme delicacy; and the carbon microphone has attained almost universal use as transmitter, on account of its great power in this capacity, and the various devices used with it to produce a strong alternating current of varying pressure.

Methods of Strengthening Magnet Receivers.—We may say that there are three ways in which inventors have sought to raise the efficiency of the magnet receiver: first, by strengthening the magnet; second, by increasing the number of the pole contacts; third, by increasing the mass of the armature.

By far the greater number of new variations fall under the first head—strengthening the magnets. This result may be

FIG. 66.—Single-pole compound bar telephone magnet.

accomplished in two ways: either by using a horseshoe magnet, both poles of which are in contact with the diaphragm; or, by "compounding" the magnet, or building it up by using a number of magnetized steel bars, in single or double-pole contact. Fig. 66 shows a compound bar magnet of the variety now most often employed in common receivers instead of the single rod at first used, as shown in Fig. 61. Compounding the magnet is a method we may compare to harnessing a number of horses to a heavy wagon, and works with the same effect of combined strength. It is also preferable to employing single bar magnets of the same degree of power; both because it is cheaper, and also because the several separate bars act on one another to prevent the magnetism from becoming exhausted as soon as it would be in any one of them used alone.

Methods of Adjusting the Receiver.—Another problem which a number of inventors and manufacturers have sought to ~lve is one of construction: how to obviate the effects of the ʼering ratios of expansion and contraction, as the weather is warm or cold, between the hard rubber of the shell and the steel of the magnet. Because this difficulty had not been met, the earlier forms of Bell instruments were often rendered useless, for the time being, by sudden changes of temperature. The first inventors to attack this problem sought to regulate the distance between the diaphragm and the pole pieces by an adjusting screw, passing through the bottom of the case so as to push the magnet forward or draw it back. In such instruments the magnet was secured at the end furthest from the diaphragm. In more modern types the adjustment is effected by securing both magnet and pole pieces to a brass plate or cup, which is part of a rigid system including the diaphragm.

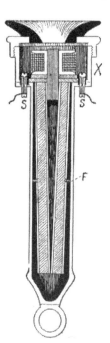

FIG 67.—Sectional diagram of the Neumayer receiver.

The Siemens Telephone Receiver.— Among the best-known types of magnet telephone is the Siemens instrument, which is widely used in Germany. In general the case, made of sheet-iron, is so constructed that it can stand upright on a table or a shelf, and this was the invariable rule with a variety in which was included also a small ringing apparatus, consisting of an armature made to rotate between the two legs of the magnet and turned by a small crank in the side of the case precisely like the ordinary magneto-generator, to be described in a later chapter. The rotation of this armature induces a current in the magnet coils causing the attraction of the diaphragm, in both transmitting and receiving instruments. A small ball laid upon the horizontal diaphragm supplies the means for making an audible signal under such stress. The magnet, made of a rather broad and thin strip of steel, is secured to the tail of the case by a screw. The pole pieces consist of two very small oblong coils brought very close together, and attached to the ends of the magnet by two thin steel plates carrying the soft iron cores. The instrument is a very powerful one, and, in practice, is frequently used as both transmitter and receiver.

The Neumayer Receiver.—Another good instrument, also a German invention, is the Neumayer receiver, a section of which is shown in Fig. 67. It combines several features not found in other instruments. The magnet consists of five cylindrical rods, which touch at the " south," or negative poles, and are

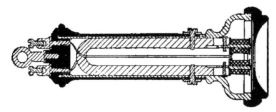

FIG. 68.—The Ericsson receiver. This instrument is manufactured in Sweden and is used by many independent companies in the United States. It is notable for the strength of its magnet and the excellent device for securing adjustment. The case is a brass tube, enlarged at one end to form the cup or box for enclosing the coils and the diaphragm, and upon this the head cap is screwed in the usual manner. The adjustment is obtained by pushing the magnet to the proper position in the tube, and firmly securing it by the screws, which work in a slot on either side, as shown. The tube of the case is covered with a coating of hard rubber. The binding-posts are held apart by a wedge of rubber secured between them by the nickel-plated ring at the tail of the case.

held in position around the core of the armature coil by the brass ring, *F*. The coil is wound on a wooden spool, and through its center runs a thin brass tube, within which is soldered the core, consisting of a number of fine iron wires, of the kind used by florists to bunch together the stems of flowers. By this device it is believed that a greater inductive action is obtained, on the principle applied in the construction of induction coils, that this " lamination " of the core reduces the effects of the induced E M F due to the rapid variations of the magnetic flux passing through it, hence overcoming the danger of induced secondary currents and lowering the magnetic resistance. With the high frequency telephonic currents this is a consideration of vital importance. The core, moreover, is in constant contact with the magnet bars, and thus in the field of magnetic force.

The coil is contained in a brass box, *X*, on the top of which is screwed the cover, clamping the diaphragm in place. The wires are attached to the binding-screws at *S* and *S*, and the magnet is contained in a wooden sheath, which is secured to the brass head box.

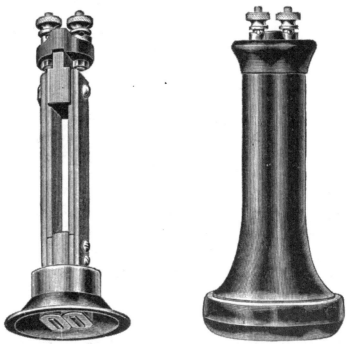

FIGS. 69 70.—Bi-polar compound bar magnet receiver of the Keystone Electric Telephone Co. The brass cup holding the magnet coils fits inside the receiver shell, being held in place when the head piece is screwed on. The magnet is in no way attached to the shell, as may be understood from the figure, thus securing a rigid and permanent adjustment to the diaphragm.

Defects of Early Receivers.—Most of the earlier types of magnet receivers had either one of two grave defects in construction: either the adjustment of the magnet was made by a screw passing up through the tail of the case, as in Fig. 62, or a screw near the poles engaged a thread in the inside of the case. In both cases the adjustment was uncertain, particularly in the hands of inexperienced persons, who would frequently damage instruments affected by sudden changes in temperature.

Boxed Coil Receivers.—In the most improved makes of instrument the brass box for containing the magnet coils, on about the same plan as that employed in the Neumayer receiver, is rapidly being adopted. It is an arrangement superior to having the armature coils in any way attached to a hard rubber shell, because, as has been ascertained, the "coefficient," or ratio of expansion or contraction, under heat or cold, is about the same for brass and steel, and an adjustment between the core of the magnet and the diaphragm, once established, is likely to be maintained under all temperatures.

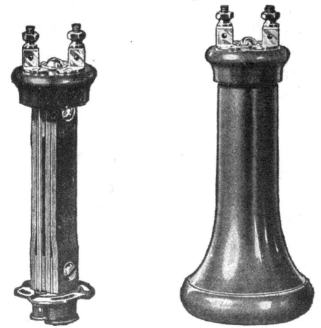

FIGS. 71–72.—Bi-polar compound bar magnet receiver of the Holtzer-Cabot Electric Co. The magnet coils rest on a metal flange, shaped as shown, which is secured to a corresponding flange in the inside of the case by four screws working in slots. Accurate adjustment may be obtained by turning the back cap, which is separate from the rest of the case, and tightening the screws at the required point in relation to the diaphragm.

Telephone Magnets.—The magnets of the better class of telephone receivers are made of the hardest tempered steel and magnetized to a degree that will enable any one of them to lift a mass of iron weighing at least four—some even as high as fifteen—pounds. In practice the general experience seems to endorse the opinion that by improving the quality of the steel the power of the magnet may be best increased.

Watchcase Receivers.—In addition to the ordinary long-magnet receiver there is another variety of a much more compact and convenient shape, which, from its general form, is called the "watchcase" receiver. One of the earliest patterns

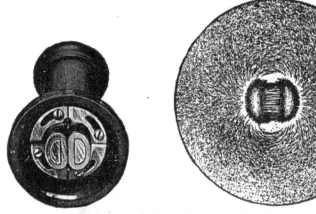

FIGS. 73-74.—Pole pieces of the Holtzer-Cabot bi-polar receiver, showing their peculiar shape, and illustration of the field of magnetic force on the diaphragm, showing the even and symmetrical pull and the utilization of all the lines of force.

of this instrument was made with a coiled magnet, on one pole of which was placed the usual pole piece, directly under the diaphragm, the whole being enclosed in a case about the size and shape of the old-fashioned "turnip" watch. Later examples of this variety of receiver have either the arrangement of the Gower telephone—a semi-circular magnet with two pole coils—or a polarized ring magnet, the two coils being attached on the diameter. Watchcase receivers are most often used, with the head-gear attachments, by telephone operators at switchboard stations.

The D'Arsonval Receiver.—An interesting variation of the magnet receiver, in which the greater part of the strength of the magnet is supposed to be directly utilized in the operation of the instrument, is the D'Arsonval receiver, widely used in France and other European countries. The inventor's contention is that the only really useful part of the field of a telephone magnet is the part situated directly between the two poles; the portion of each coil outside of this area being almost completely lost, and only furnishes a useless resistance to the current.

FIG. 75.—Interior view of a typical "watchcase" receiver.

With this theory in mind, he constructed the instrument shown in Fig. 78. The permanent magnet is spiral-shaped, one pole ending directly under the center of the diaphragm and bearing the core of the pole-coil; the other entering at the side of the wooden receiver case, and having at its end an iron cylinder which completely envelopes the coil, thus surrounding it completely in a magnetic field of great intensity. The diaphragm is clamped in place in the usual fashion, and the conducting cords enter at the side of the case opposite the negative pole, connections being made inside. The instrument has given excellent results, although it has failed of a very wide use.

The Phelps Crown Receivers.—The multiple contact receivers are less numerous, although worthy a passing notice. The earliest types were the Phelps "crown" receivers, designed

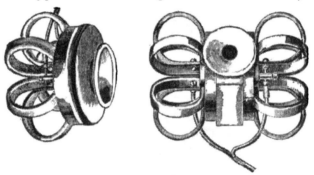

FIGS. 76-77.—Phelps single and double "crown" receivers.

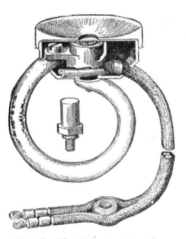

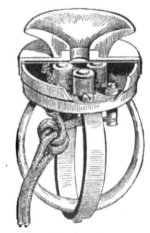

FIG. 78.—The D'Arsonval receiver.   FIG. 79.—The Goloubitzky receiver.

in 1878. The "single crown" type is shown in Fig. 76. Here the magnets, bent double, are arranged around a common center, the north, or positive, pole of each one carrying a separate coil, and the negative poles ending at the rim of the diaphragm. The "double crown" receiver, shown in Fig. 77, is somewhat

different in construction, being really two single crown instruments, with two diaphragms separated by a common vocalizing
chamber, in which was also introduced a cone-shaped piece of
non-conducting material. Both these instruments gave good
results and loud, distinct articulation, but the expense of making
them prevented their general adoption.

The Goloubitzky Receiver.—An instrument of somewhat
similar appearance is the Goloubitzky receiver, as shown in Fig.
79. It consists of two circular magnets, to the four poles of
which are attached the usual cores and coils; these being wound

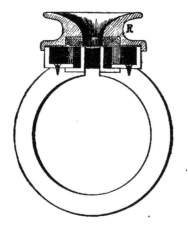

FIG. 80.—The Ader receiver.   *R*, the super-exciter.

in series, or the wire from the positive coil running over to, and
around, the next negative coil, so that there is a negative contact at one terminal and a positive at the other. Goloubitzky
observed that when the same message was delivered to several
receivers at a station at once the sound was equally loud and
distinct in each; consequently, he reasoned, that to increase the
number of contacts in one instrument—really combining two or
more instruments in one—would increase the power of that
instrument. The results, however, although good, hardly warranted the added expense and care required to get perfect
receivers of this type.

The Ader Receiver.—The last theory of improving the magnet telephone is that of strengthening the armature. The best known type of this construction is the Ader receiver, as shown in Fig. 80. It consists of a single magnet, bent into almost circular form, and carrying a coil on each pole, after the fashion of all double-pole instruments. The magnet is generally nickel-plated, to serve as a handle. The special feature of this receiver is the small ring of iron, *R*, encircling the inner end of the mouthpiece, as shown. This ring, which is called the "super-exciter," acts to render the field of magnetic force more intense, and is doubtless one cause of the remarkable efficiency of the receiver. It has been remarked by several telephone authorities that the intensifying effect would be increased, in both this instrument and the D'Arsonval, if the whole mouthpiece were made of soft iron.

FIG. 81.—The cord of a telephone receiver. It consists of two strands of braided tinsel, insulated, and enclosed in a woven sheath. A metal pin is secured to the end of each strand, and this is inserted in the grooves of the binding-post to give electrical connection.

There is a large number of receivers of patterns different from those described, in one point or another, but all combine the essential features of a permanent magnet, or magnets, bearing coils on the poles, and an iron diaphragm to receive and turn back into sound-waves the impulses brought along the line from the transmitting instrument at the far-away station occupied by the party with whom the conversation is being held.

The instruments described are almost all in practical use, or have been, and an understanding of their theories gives a fair idea of the progress of telephony in several countries.

CHAPTER SEVEN.

THE CARBON MICROPHONE TRANSMITTER.

Conditions for Good Transmission.—As we have seen, the earliest form of the Bell telephone circuit was simply two lines of wire ending at the poles of two magnet instruments, with no battery. This arrangement is very satisfactory for short lines, where there is little or no interference from other electrical circuits, or the danger of meeting with unexpected objects that will impede or deflect the current. To meet such obstacles, and at the same time transmit a clear and distinct message, a stronger current is needed than the magnetic field of a Bell instrument can generate. As we have seen, also, the effect of an alternating current and varying resistance is another necessity for getting distinct utterances at a distance.

The Theory of Varying Pressure.—The need of a varying pressure to perfectly modulate the current was met in the Reis transmitter, but the light contact between the electrodes, as we have described it, created the constant danger of breaking the circuit and thus interfering with the perfect transmission of sounds. Conseqently, shortly after the introduction of the magnet telephone, a number of inventors began working on the problem of a practical transmitter, which should embody the principle set forth by the French scientist, Du Moncel, that, "if the pressure between two conducting bodies forming part of an electric circuit be increased the total resistance of the path between them will be diminished, and if the pressure be decreased, there will be an increase in the resistance."

Berliner's Transmitter.—One of the earliest attempts to make a variant pressure transmitter is shown in the instrument invented by Emile Berliner in 1877. It consisted of a vibrating metal diaphragm of exactly the kind used in the magnet receiver, and against the center of this was a metal knob, or button, held in place by a thumb-screw. One electrode was attached to the diaphragm and the other to the metal ball. By speaking against the diaphragm we cause it to vibrate, and hence alternately press hard and gently on the button, thus varying the pressure of the current supplied by a battery. Its working was not satisfactory.

Edison's Carbon Disc Transmitter.—After an extended course of experiments to determine the most suitable substance

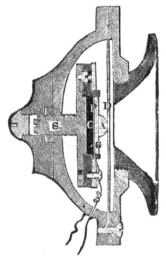

FIG. 82.—Edison's Carbon Disc Transmitter.

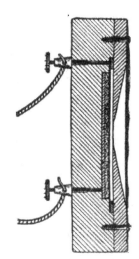

FIG. 83.—Hunnings' Carbon Dust Transmitter.

for a varying pressure transmitter, Mr. Edison designed the carbon disc instrument, which embodies the theory on which transmitters have been constructed up to the present time. His first pattern instrument consisted simply of a button of plumbago, held against a small platinum disc and secured to the diaphragm by an adjusting screw. Later he modified his plan, and designed the instrument shown in Fig. 82. Here the carbon button, *C*, made of compressed lampblack, is placed between two discs of platinum foil, *F*, and attached to the enlarged end of the adjusting screw, *S*. In front of the forward disc of platinum rests a plate of glass, *G*, to which is attached a rounded button of bone or ivory, *A*, resting against the center of the diaphragm, *D*. This was a more practical instrument than was Berliner's, but, as was later shown by Prof. David B. Hughes, of England, the carbon transmitter requires, not only a variant pressure, but, also, a loose contact of the electrodes at th : start. In the theory of loose contact we see a memory of Reis's instrument.

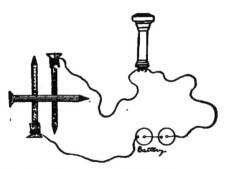

FIG. 84.—Diagram of the Hughes Micr)phone.

Hughes's Microphone.—Hughes based his theory on a device invented by him, known as the microphone, the simplest form of which is made as shown in Fig. 84. Here we have a circuit of wire, including a voltaic cell and a telephone receiver, and having as terminals two steel wire nails. The circuit is closed by laying a third nail across the other two. If, now, a sound, as of the ticking of a watch, be made near the nails, it may be distinctly heard in the telephone receiver, although a considerable distance intervene. A more perfect form of microphone made on the same principle, consists of a pencil of battery carbon, sharpened at both ends, and loosely attached in sockets hollowed in two carbon blocks, which form the electrodes of the instrument. The whole, mounted on a resonant sounding-board, is extremely sensitive to minute sounds— being literally, as its name indicates, to small sounds what the microscope is to small objects.

Hunnings' Dust Transmitter.—Following up this theory of loose contact, Henry Hunnings, in 1881, devised the original carbon grain transmitter, thus making a distinct departure in construction and giving the pattern for what has become the prevailing form of instrument, particularly in America.   Fig. 83 shows a sectional view of this instrument.   It is a circular wooden block in which a shallow cavity has been hollowed out.   In this cavity is set a metal disc, making one electrode connection, and over it are a number of carbon granules, generally coke ; the whole being capped by the metal diaphragm, which forms the other electrode, and is clamped in place by the mouthpiece.   The effect of the numerous granules of carbon is to give a multitude of contacts to transmit the varying stress of the diaphragm by the electric current.   The result, in point of clear and loud transmission, is similarly increased, a fact which has caused the granule or "dust" system to be so widely adopted. The principal trouble, however, is that, in the simpler forms of this pattern, the granules are apt to pack under the oft-repeated stress on the diaphragm, and the efficiency of the instrument is thus impaired.

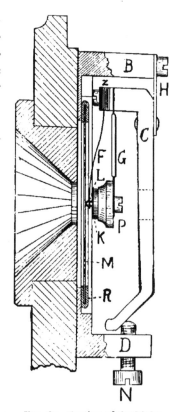

FIG. 85.—Section of the Blake Transmitter.

The Blake Transmitter.—Another instrument that has seen considerable use both in this country and England is the Blake transmitter, shown in Fig. 85.   Here, *M* is the diaphragm, behind which is screwed an iron ring, carrying the two projecting pieces, *B* and *D*. Upon the upper projection, *B*, is fixed the angle-piece, *C*, by means of the flexible brass strip, *H*, with its lower angle abutting against the adjusting screw, *N*, which is fitted into *D*. It is by this screw that the instrument may be adjusted whenever its working becomes defective. The diaphragm, an iron disc, is surrounded by a rubber ring, *R*, which is secured behind the mouthpiece by two "damping" springs, one of which presses on the rubber ring and the other on the diaphragm. On the upper arm of the angle-piece, *C*, are fixed two springs, *F* and *G*, the former of which is insulated from the metal of *C* by a clamp of hard rubber, or other suitable non-conducting material, *Z*. The latter, *G*, is in direct electrical connection with the metal of *C*. The spring, *F*, carries on its extremity a short section of platinum wire, *L*, which rests directly against the center of the diaphragm, on one side, and against the carbon button, *K*, fastened in the brass cup, *P*, at the end of the spring, *G*. The motion of the diaphragm varies the pressure between the platinum point, *L*, and the carbon button, *K*, and thus varies the current which passes from the battery through the iron frame ring, thence through *B, H, C, G, P, K, L, F*, and out again from the contact, *Z*, of the spring, *F*. The rather complicated series of springs and levers is intended to hold the contact pieces in position, so that no variation of the sound may be lost, either through a too loose contact or the wearing of the carbon disc.

Varieties of Carbon Transmitters.—Both the Hunnings and the Blake types of transmitter are constructed on the principle of the microphone—both have one or more carbon electrodes in loose contact, susceptible of varying pressure, according to the movements of a thin metal disc, moved by the stress of the voice. There are three general types of carbon transmitter, all embodying these same principles: 1. Carbon pencil transmitters. 2. Carbon granule transmitters. 3. Ball or button transmitters. Of these three kinds, the first is most like in pattern to the original Hughes microphone, consisting of two or more carbon pencils arranged in carbon sockets in a variety of orders around the center of a wooden diaphragm. The idea of thus varying the arrangement is that the sound impulses, striking the diaphragm, spread in various directions and should be met by appropriately arranged carbon pencils.

The D'Arsonval transmitter is among the most elaborate, each pencil being covered with a cylinder of sheet iron and a permanent horseshoe magnet being mounted behind the row of pencils, to provide an added means for producing a variable pressure, as it attracts the iron covers and holds them so long

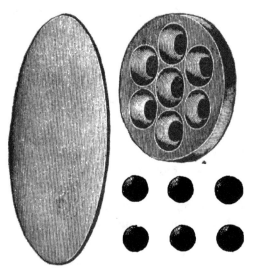

FIG. 86.—Carbon Diaphragm, Back Electrode Plate and Carbon Balls for the Ball type of Transmitter.

as the stress of the voice on the diaphragm continues the altered pressure. Such transmitters are common in France and other European countries, but have gained no favor in America.

Carbon Ball Transmitters.—The ball, or button, transmitters are equally interesting and various, although the principle of their construction is only a type of the same kind of instrument as the pencil transmitter. Fig. 86 shows a form of this instrument, which consists of three parts: a carbon diaphragm, a carbon block, or plate, having a number of round

cavities, and the same number of carbon balls, of a size to fit
loosely into the cavities and be held in place when the carbon
plate carrying them is fastened into a frame with the diaphragm.
Another form of this instrument has one large ball, fitting into
a cup of carbon or metal and held against a carbon diaphragm
in the same manner.

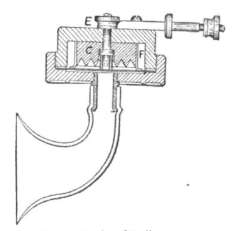

FIG. 87.—Section of Berliner's   FIG. 88.—Back Electrode Carbon Plate
    Universal Transmitter.   for Berliner's Transmitter.

Berliner's Universal Transmitter.—Several types of modern dust transmitters employ, as the back electrode, a block of
carbon with a series of cones or concentric circles, as shown in
Fig. 88.   The plate, thus worked, is separated from a carbon
or metal diaphragm by a quantity of carbon granules, which are
evenly distributed in the spaces between the cones or circles,
thus preventing packing at any one point and distributing the
pressure.   Of this type is the well-known Berliner Universal
Transmitter, depicted in Fig. 87.   Here we see the peculiarities
of its construction, which renders it desirable to place the carbon diaphragm in a horizontal position and allow the back electrode, the circular-grooved carbon plate, to rest upon it.   The
grooved plate, C, surrounded by a cover of felt, F, which
touches the diaphragm about its circumference, is held just
above the diaphragm by a quantity of carbon granules to give

the variable resistance. On the center of the diaphragm rests a short rubber tube, held in place by the micrometer screw, *E*, thus serving to damp or modify the vibrations of the diaphragm. The same screw holds the carbon block in position and also holds the back electrode. The other electrode is at the brass ring which clamps the carbon diaphragm into place.

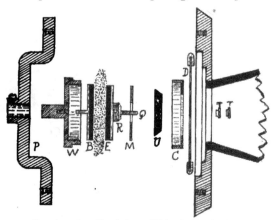

Fig. 89.—Details of the Solid-back Transmitter.

The Solid Back Transmitter.—A form of carbon transmitter most often used in the United States is shown in Fig. 89. It is called the "solid back," or White transmitter, and is one of the most efficient in use. The back electrode is a metal case, *W*, secured by an adjusting screw to the supporting bracket, *P*. It is lined on the inside with white paper, to serve for insulation. Inside, against the rear wall, fits the small brass disc, *B*, which carries a pin on its center to attach it to the support, and on its face carries a button or plate of carbon. In front of and around this carbon disc is a quantity of granules, which rest on the other side against another carbon disc, carried on the brass button, *E*, from whose center projects the screw-threaded bars, *R* and *Q*. Over these fit the mica washer, *M*, the nut, *U*, and the cap, *C*, which is screwed in position. The bar, *Q*, projects through the center of the vibrating diaphragm, *D*, and is held in position by the nuts, *T* and *T*, which act as dampers.

Improved Transmitters.—The number of new transmitters constantly being put upon the market by various manufacturers is simply immense. Many of the features embodied are undoubtedly excellent, and many marked improvements have been made, both in efficiency and durability. Where carbon grain is used the principal problem has been how to prevent packing, or solidifying, which inevitably results in disabling the transmitter by destroying the loose contact.

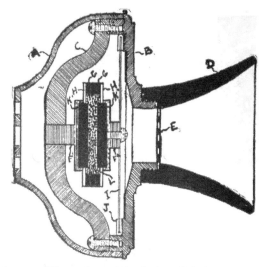

FIG. 90.—Improved Transmitter of the Century Telephone Construction Co.

The Century Transmitter.—The improved transmitter of the Century Telephone Construction Co., shown in Fig. 90, presents many points of excellence that are worthy consideration. It is one of the best of the several improved forms of the "solid-back" transmitter, just described. Here, *D* is the hard rubber mouthpiece; *A*, the cup containing the electrodes; *B*, the cover, secured to the bridge-piece, *C*, by the screws, as shown. The extra strength of the bridge-piece is one of the features of this instrument, securing, as it does, perfect rigidity, and obviating all danger of warping or "buckling," as must frequently occur in ill-constructed instruments. In the center of the bridge is screwed the brass cup, F^1, which contains a button of compressed carbon, K^1, and forms the back electrode. A similar cup, F^2, is screwed to the diaphragm, as shown, and forms the front electrode. The two are separated by the mica washers, H and H, and the felt washer, G, which form a chamber, L, to be filled with carbon granules. As will be seen, the chamber thus formed is enlarged into pockets at top and bottom, and into the lower one the carbon granules, having a tendency to pack, will settle out of the path of the current. Rapid and accurate adjustment may be obtained by turning the screw at F^1.

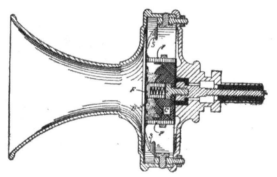

FIG. 91.—Ericsson Transmitter.

The Ericsson Transmitter.—Another form of transmitter which is winning wide favor is shown in Fig. 91. It is manufactured in Sweden, and imported by the Ericsson Telephone Co., of New York City. The principal features claimed for it are high efficiency and distinct transmission, with small battery power. The carbon granules used are prepared by a secret process—a fact that has enabled the instrument to maintain its high reputation for many years. Reference to the figure will show that the general construction is much like that of the Berliner Universal Transmitter, already described. Here, D is the diaphragm of ferrotype iron, to the rear side of which is secured a gold-plated disc, forming the front electrode. To the outer face of the diaphragm is cemented a cover of oiled silk, which proves an effectual protection from moisture. The grooved carbon block, *C*, forming the second electrode, is secured to the rear of the case, from which it is properly insulated, and connection is made by the screw protruding at the end. Two double damping-springs, *S* and *S*, hold the diaphragm in place. A ring of felt, *FF*, surrounds the rear carbon block, and, bearing against the diaphragm, serves to retain the carbon granules. Another ring of felt at the center of the block surrounds a small coiled spring, which still further damps the diaphragm, and prevents harshness of tone in speaking.

FIG. 91*a*.—Hand Microtelephone Desk Set, with Magneto Generator and Bell Mounted in Base.

CHAPTER EIGHT.

THE CIRCUITS OF A TELEPHONE APPARATUS.

Electrical Source of the Transmitter.—As we have seen, the Bell magnet telephone, and the numerous variations of it, operates by a current produced by the agitation of a magnetic field, somewhat after the manner of a dynamo. We have also learned that, while it is immensely sensitive to small currents, it cannot produce currents of sufficient pressure to successfully operate on long lines of wire, an alternating current of high voltage and variant pressure being best adapted to transmit the sounds of the voice and produce effects on the diaphragm of the receiving instrument. In the various types of transmitters we have described, we have seen that the apparatus for producing the variation of pressure is composed of two carbon contacts, or one carbon and one metal contact, but in this arrangement there is no provision for producing a current, as in the magnet telephone. The transmitter can only vary a current, consequently there must be some other source of electrical activity. This is supplied by a commercial voltaic cell, or battery of several cells, and in general usage we have the type of cell known as the "open circuit."

Open Circuit Batteries.—The open circuit cell is so called because it has the power of recuperating its strength, or "depolarizing," whenever the circuit is left open or is not in use. The closed circuit cell works best when in constant use, and has the power of constantly depolarizing itself. It is used where a strong current is constantly needed, as in telegraphy.

The condition known as "polarization" is, briefly, due to the collection of minute bubbles of hydrogen gas on the face of the negative plate, which interferes with the production of electromotive force by reducing the active surface of the plate, and

hence increasing the internal resistance of the cell. This effect
is remedied in closed circuit batteries by a substance, known as
the "depolarizer," which in process of electrolysis constantly
releases oxygen gas to combine with the free hydrogen. In the
ordinary blue-stone Daniell cell the depolarizing agent is the
sulphate of copper ("blue vitriol"), which constantly releases
oxygen and causes a deposit of metallic copper on the face of
the copper electrode. The process of depolarization is, thus,

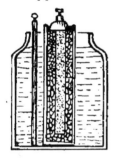

FIG. 92.—Section of ordinary Leclanché cell.

dependent on the continuity of the current,
and ceases with it. For this, among other
reasons, this type of cell rapidly deteriorates
when not in use. In the open-circuit cell,
on the contrary, the most common and
efficient type of which is the Leclanché,
so named from its inventor, a celebrated
French electrician, the depolarizer is dioxide of manganese, which surrounds the
carbon plate When the circuit is open and
current has ceased to flow, the oxygen is
constantly given off from this substance to combine with the
free hydrogen, and the manganese then absorbs a fresh supply
from the air. Thus the battery is in best condition after a
season of rest, and is not capable of continuous work, like the
cells of the other type. This is the reason that open-circuit
cells are used in telephone practice on short lines, although in
long-distance telephony, where strong currents are required for
long periods, the closed circuit cells are coming into more
extended use. Fig. 92 shows a section of a common type of
the Leclanché cell. In this figure the carbon plate is inserted
in a porous cup and surrounded with a quantity of granulated
carbon and manganese oxide. The zinc electrode is a pencil at
the side of the porous cup, both being immersed in a solution
of sal ammoniac.

Dry Cell Batteries.—Instead of the type of cell just described "dry cells" are frequently used in connection with the

station telephone apparatus. These are in all respects like the ordinary cells, except that the liquid solutions are rendered practically solid by being mixed with such substances as zinc oxychloride, gelatinous silica, or powdered gypsum. The advantage of this arrangement is that there are no liquids to spill and cause damage.

Strength of Current Required.—According to the principle that the variation in the strength of the current passing through the transmitter is the most important thing next to the consideration that the resistance should be as low as possible, so that the slightest variation should be effective, it would seem to follow that the stronger the current the more efficient the transmission of the voice. This is, however, not the case, as after a certain point the current tends to heat the carbon contacts, and thus destroy the transmitter in time. For this reason, a low resistance, low voltage battery, like the Leclanché, is the best for most short-distance transmission. In long-distance work there is needed a special form of transmitter if the use of the line is constant. Then a strong constant-current battery may be placed in circuit.

The Induction Coil.—During the first five years of practical telephony (1876–81), the transmitter and receiver were always placed in one circuit. This was also true after the carbon transmitter had come into general use. In practice, however, it was soon found that practical results could not be obtained with the instruments in direct connection, since the changes produced in the total line resistance by the varying pressure of the transmitter electrodes were so minute in comparison as to be scarcely perceptible in the receiver at the other terminal of the circuit. In 1881 Mr. Edison devised a most effective method for remedying this difficulty ; connecting the two terminals of the circuit, including the transmitter and galvanic cell, with the primary winding of the induction coil, the secondary winding of which was connected direct to line. By this arrangement of the circuits—which was really Professor Gray's idea, and had

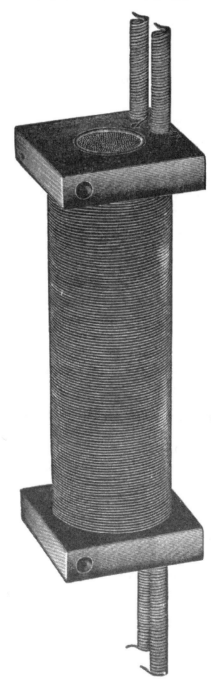

FIG. 93.—A telephone induction coil of usual size, showing the laminated core and the attachments of the primary and secondary windings.

been used by him in connection with his harmonic telegraph as early as 1875—not only was the pressure of the line current increased in a ratio of about ten to one, but a rapidly alternating current was obtained in place of the direct, varying pressure one previously used.

Action of the Induction Coil.—As we have already seen in the section on induction, the inductive influence of one circuit on another becomes perceptible on two occasions : either when the circuit is made or closed, and again when it is broken; or, in the same way, when the strength of the current is increased, either by adding to the voltage or diminishing the resistance, and again when it is similarly diminished. Thus each distinct sound, as it affects the diaphragm of the transmitter, decreases the resistance in the primary circuit of the induction coil, and induces a momentary current in the secondary winding in the opposite direction to that in which itself is moving. Each successive reaction of the diaphragm and the variable pressure electrodes at the completion of every successive sound induces a momentary current in the same direction as that in which the primary current is moving. Thus the full effect of current alternation is obtained.

Size of Telephone Coils.—The dimensions of telephone induction coils, suitable for given instruments, is a matter of considerable calculation and experiment, in which questions regarding the size of the primary and secondary winding-wires and the length of the line must be carefully considered. Coils for long-distance work are longer and of greater inductive capacity than those used on shorter lines. One of the most efficient makes of induction coil for ordinary service is constructed as follows: The core is composed of about 500 lengths of wire of size 24, B. & S. (American Wire Gauge, or Brown & Sharpe), each wire in the bundle having a diameter of about two one-hundredths of an inch, or 20.100 mils. The bobbin is four inches in length, and the core nine-sixteenths inch in diameter. The primary winding consists of about 200 turns of No. 20

wire, silk-covered, the diameter being about three one-hundredths of an inch, or 31.961 mils, and is two layers deep. The secondary winding consists of about 1,400 double turns (two wires being wound side by side) of No. 34 wire, the diameter of which is about six one-thousandths of an inch, or 6.403 mils. In a circuit including this coil it has been found that the primary winding has a resistance of thirty-eight one-hundredths of an ohm, and the secondary about seventy-five ohms. It is necessary that the resistance of the primary circuit be as low as possible, in order that the minutest change in the resistance at the carbon electrodes may exert the greatest possible change in its circuit. The effect of the induction coil is to increase the electromotive force going upon the line in about the ratio that exists between the turns of the two windings. Thus, if the primary winding consists of 100 turns, and the secondary of 2,500, the ratio in electromotive force between the two will be about 1 to 25. With this increase in the pressure comes a corresponding increase in the strength and amplitude of the impulses generated in the transmitter, by the variation of the stress at the carbon contacts. This increase in the pressure of the current on line permits the transmission of the same impulses to the receiving station.

An Alternating Transmitter.—Several inventors have occupied themselves with the problem of making a transmitter that shall produce a true alternating current independent of the undulations of pressure that produce this effect through an induction coil. One of the most successful instruments of this type is the one invented by G. F. Payne and William D. Gharky, two telephonists of Philadelphia. Fig. 94 shows a section of this instrument, in which *A* represents the mouthpiece of the transmitter, enclosing one side of the diaphragm, *D*. *B* represents a closed cylindrical box, within which are five piston-shaped electrodes, three of them, *X*, *Y*, *Z*, being fixed, and two others, *M* and *N*, arranged to slide back and forth as impelled by a piston-rod attached to the center of the diaphragm by a screw piercing its center. Each electrode bears plates of carbon, shown at *C, C, C, C, C.* The spaces between the electrodes are filled with granular carbon, which is prevented from leaking out into other compartments by the felt washers, *F, F, F, F.* The battery wires are so attached that the movable electrodes, *M* and *N*, form the two terminals of the circuit. The three stationary electrodes, *X, Y, Z,* are connected, as shown, with the primary winding of the induction coil.

The method of operation is as follows: When the sounds of the voice strike the diaphragm they produce the changes characteristic of sound waves, causing the diaphragm and the piston attached to it to vibrate, first inward and then outward.

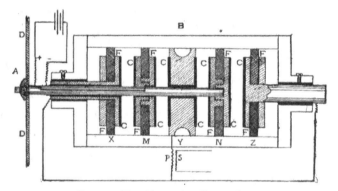

FIG. 94.—The Alternating Transmitter.

As a result of this action and reaction, in varying degrees of force, the electrodes *M* and *N* are first brought into electrical contact with the electrodes *Y* and *Z*, and then, with *X* and *Y*. Thus, by the first motion, the circuit is closed by the wires at the right and center, leading into the primary of the induction coil; by the second motion, by wires at the left and center. So the center wire is alternately positive and negative, and the left and right ones, in turn, form the opposite poles of the circuit, and the current passing through the primary of the induction coil, flows first in one way and then in the other. The effects of alternation are thus vastly increased

The Magneto-Generator Bell Call.—Before explaining
the circuit arrangements of a telephone apparatus, it will be
necessary to deal with just one more contrivance—the bell call.
As we have already seen, this consists of two gongs that are rung
by the current generated in a magneto-electrical machine by
turning a crank handle at the side of the box mounted at the
top of the telephone back board.   As constructed by the various
manufacturers, this instrument consists of two, three, four, or
even five, large horseshoe permanent magnets, so arranged that
an armature mounted on the spindle of the crank can turn

Fig. 95.—Core of a magneto-generator armature, and end view of same.   The insulated
wire is wound on from end to end, parallel to the spindle and almost to the top of the
flanges at either side, which are left bare.   The form of core here shown is "laminated,"
or composed of a number of pieces, as indicated by lines running across the axis.   This
improvement was introduced by the Holtzer-Cabot Electric Co., and has the advantage
of permitting a more accurate adjustment, a greatly enlarged field, and consequent
increase in output of power.

between the poles of the magnets.   It will be seen that it is a
dynamo-electrical machine in all respects except that permanent
magnets are used instead of electro-magnets.   Fig. 95 shows
the armature core and Fig. 96 the method of winding on the
insulated wire, and the position of the armature in the pole
casting.   This wire, thus wound, acts to all purposes like the
single loop armature shown in Fig. 28, except that the large
number of its turns permits a variation in the number of the
lines of magnetic force that pass through it in its various positions.   When it is horizontal these are fewest; when vertical,
in the greatest number.   On this account, when the movement
is from the horizontal to the vertical, the current generated flows
through the coil in one direction, say, from right to left, as the
generator stands in its box in the telephone apparatus; when
the movement is from vertical to horizontal, the number of mag-

netic lines constantly decreasing, the current flows from left to right through the coils. The magneto-generator, from its very construction, is, accordingly, an alternating current dynamo.

Strength of the Current Generated by the Magneto Call Generator.

—The current generated is a high pressure one, sufficient to carry the electrical impulses along an extended line wire, and ring the signal gongs at the distant station, or operate

FIG. 96.—Details of a magneto-generator, showing the armature in position and a portion of the winding. The features of this machine, which is manufactured by the Connecticut Telephone and Electric Co., are the placing of the gear and pinion wheels opposite the crank, which prevents "grinding" when the crank bearing becomes worn ; also, the perforation of the armature casting, bringing the magnets directly to the face of the armature and allowing very efficient magnetic action.

the switchboard drop, in the manner to be subsequently explained. Although on short private lines the ordinary voltaic cell and electric bell, such as is used in houses, is attached to the telephone apparatus, an arrangement of this kind is altogether too weak to suit the needs of commercial, central station lines. On these lines, as in the "bridging" system of mounting telephones, the apparatus must often ring through a resistance of several thousand ohms, sometimes 10,000 ohms, which is a point of pressure far beyond the capacity of the strongest battery of voltaic cells practicable in connection with telephone systems of the usual type.

The Bell Magnet.—The current generated thus rings the bell, in both the calling and the receiving apparatus, by energizing the coils of an electro-magnet of the double-pole type, and causing it to attract its armature, to which is attached the clapper of the bell gongs. The construction of this device is shown in Figs. 97-98. As will be seen, the rod of the ringing clapper is attached to a bar of iron which is pivoted at the center. The pivot-pin is inserted in a piece which is fixed parallel to the axes of the magnet coils, so as to allow the armature to sway from side to side as it is attracted, first by one pole and then by the other. This pivot-post is an U-shaped permanent magnet,

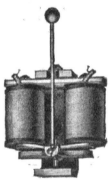

Fig. 97.—Polarized ringer magnets and clapper. View from beneath.

Fig. 98.—Polarized magnets and bells. View from above, showing manner of attachment.

and its function in the apparatus is to act as a "polarizer." That is to say, its duty is to magnetize by induction the pivoted armature and the cores of the electro-magnet. Thus, if the two cores of the magnet coils acquire a polarity of a positive quality, the armature will be negative, and, as a result, will be attracted to the one pole or the other of the electro-magnet. Since the coils are oppositely wound, a current passing through them will tend to strengthen one pole and weaken the other; hence causing the armature to sway toward the pole of the greatest strength. Then, because the current produced in the generator is an alternating one, flowing in one direction and then in another, the armature is attracted by each pole successively, with the result that the rod, or clapper, vibrates with great rapidity, striking first one gong and then the other, according as the currents cause the armature to be attracted to the poles.

FIG. 99.—Extension bells for telephone apparatus. These are constructed on the plan just described, but are connected to the apparatus by wires, so that they may be placed at any distance in order to insure the answering of a call that, otherwise, might not be heard.

Resistance of the Generator.—The armature of an ordinary hand generator is usually wound to an internal resistance of at least 300 ohms, and in some cases as high as 700. The resistance of the common type of ringer coil is between 75 and 100 ohms, although, when intended for use in bridging instruments, it is sometimes wound as high as 1,000 ohms. The purpose of inserting this high resistance, and impedence, in circuit, will be explained later. The telephone ringing apparatus is one of the most efficient instruments of its kind in the market, and the delicate adjustment of its parts, the result of years of practical experiment, enables the telephone to be the useful device it is, providing a thoroughly practical calling apparatus for even the longest lines and the noisiest stations.

In order to avoid a rather natural misapprehension, it would be well to mention here that when, in telephone parlance, one speaks of a generator of so many thousand ohms, reference is not made to the internal resistance of the armature winding, but to the output-power in $E M F$ of the generator. Thus by a generator of 50,000 ohms we mean one that can ring its own bell through a line of that resistance,

CHAPTER NINE.

THE SWITCH HOOK AND ITS FUNCTION IN THE TELEPHONE APPARATUS.

The Automatic Cut-out.—From the descriptions so far given it may be readily seen that in a telephone apparatus there are two distinct circuits—the calling circuit and the speaking circuit. Even a novice can understand that both cannot be included in the line at one time; since it is evident that the great resistance to the current offered by the generator and bell magnet coils would materially interfere with the successful transmission of the voice. Accordingly, there is a perfect system by which either of these apparatus is cut out of circuit while the other is in use. The result is accomplished by the device known as the switch hook, the working of which has been described in Chapter One.

The Attachment of the Switch Hook.—One form of hook switch is shown in Fig. 100. While differing in some details of construction from the types produced by other manufacturers, it possesses all the essential features we need to understand. These are, briefly, the three points of electrical contact—two below and one above the shank of the hook lever. As the hook in this cut is up—that is, relieved of the weight of the receiver, which, as we have seen, is intended to hang upon it—we see that the contact of the shank is with the two springs below. When the hook is down, or has the receiver hanging on it, the contact with the lower terminals is broken, and connection is made with the upper one. This is the position of the switch hook when the telephone apparatus is not in use, and the receiver is hung. Then, by the arrangement of the wires ending at the lower and the upper contacts, the transmitter and receiver are cut out of circuit, and the calling apparatus is connected direct to line. Similarly when the receiver is removed from the hook, and it is allowed to spring up by the tension of the leaf spring which bears on the shank, as shown in the cut, the circuit of the transmitter battery is made, and messages spoken into the transmitter at another station may be heard in the receiver. This is the reason why the crank of the call-bell generator is always turned before the receiver is lifted from the

FIG. 100.—One form of switch hook, showing single electrical contact for the shank, when in depressed position, and double contact, in operation, when it is raised as in the cut.

hook. After one has taken down the receiver it is useless to continue ringing the call bell, as he merely makes a noise in his own office without in the slightest degree attracting the attention of the central station, or of the man at the other end of the line.

"Hook Down" and "Hook Up."—Fig. 102 shows, in diagram, how the switch hook operates to open and close the circuits of the telephone apparatus. The first section of the figure shows the conditions at "hook down"; the second section at "hook up." The dotted lines indicate the wires not in use on either occasion. As may be understood, with very little study, a current entering the apparatus along the line wire, when the hook is down, will pass through or around the magneto-generator, by means of an automatic "shunt," to be presently explained; thence through the coils of the call bells, causing them to ring; afterward along the wire to the lower contact of the switch hook, through the shank and fulcrum of the hook lever, and out by the return line wire, in case it is a metallic circuit, or to ground, in case it is a grounded return circuit. As soon as the hook is allowed to rise, by the removal of the receiver, the circuit of the magneto and call bell is broken, and that of the talking instruments thrown in, by the

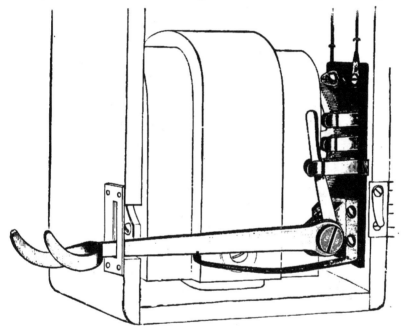

FIG. 101.—Long shank hook lever with "knife switch" attachment. Lever is shown in depressed position.

contact of the hook shank with the two terminals, as shown in the cut. Then a spoken message entering the apparatus along the line wire, as before, passes through the coil of the receiver, to be delivered at the diaphragm ; or, a message spoken against the diaphragm of the transmitter, passes out by the same wire, by means of the primary and secondary windings of the induction coil, through the two contacts of the hook switch.

Generator Cut-outs and Shunts.—In the practical working of a telephone line it would be extremely undesirable to allow the current entering the apparatus, for the purpose of sounding the call bell, to pass through the high resistance coil of the magneto-generator. Such a thing would greatly decrease the power of the current to ring the call bells. Thus, while it is desirable to have the circuits of the magneto so arranged that it can be readily thrown in, it is equally necessary to have

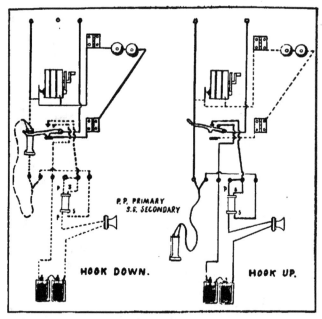

FIG. 102.—Diagram Illustrating the Manner of Changing Circuits of the a Telephone by Lowering or Raising the Switch Hook Lever.

it shunted, or bridged out when not in use. To accomplish this result a number of ingenious devices have been adopted, and are manufactured in connection with the apparatus of the several makers of telephones. The object of all is to furnish a line of lower resistance than the coil of the generator, and, by this means, to shunt the incoming current around the armature, and thence to the magnets of the call bells. ·

Familiar Shunting Devices.—Two of the most typical forms of automatic cut-out are shown in Figs. 103 and 104. Fig. 103 shows the form of shunt used by the Bell Telephone

Company. Its construction and theory are simple and effective.
The gear wheel, *A*, is mounted on the crank-shaft, *B*, in such
a way as to allow it some small freedom in turning. The shaft,
B, bears a spiral spring, *C*, which is held against the terminal
post, *D*, by the binding collar, *E*, the result being that the
point, *F*, of the crank is held in contact with the leaf-spring, *G*.

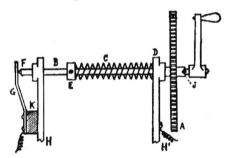

FIG. 103.—Armature Shunt of the Western Electric Co.

Thus an electric current entering the apparatus through the
wire, *H*, passes through the spring, *G*, along the shaft, *B*, and
thence to the coils of the call bell through the wire, *H'*. This is

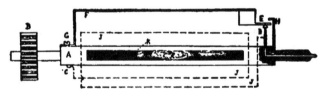

FIG. 104.—Post Centrifugal Armature Shunt. J, J, J, Diagram of armature winding;
K, End of the core flange, as shown in Fig. 95.

the arrangement which exists when the magneto is out of use, and
the call bell free to be rung by currents coming from without.
So soon as it is desired to operate the generator, this shunt is
broken by the mere act of turning the crank handle. Turning
the handle moves the pin, *J*, out of its slot in the hub of the
gear wheel, causing the binding collar, *E*, to compress the
spring, *C*, against the post, *D*, and thus to release the point, *F*,
from its contact with the spring, *G*. Because the spring, *G*, is

attached to the terminal post by an insulating block, *K*, the circuit is broken through it, and the coil of the generator is thrown in. This cut-out is a true shunt, affording a path of the lowest resistance in place of the long fine wire of the armature coil.

The Post Cut-out.—Fig. 104 shows the Post cut-out, so-called from its designer. Its operation depends upon centrifugal force instead of a pull spring. Here *A* is the shaft of the armature shuttle, which is turned by the pinion, *B*, worked by the gear wheel of the crank shaft. One end of the armature

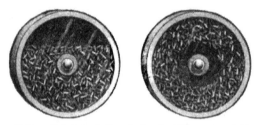

Fig. 105.—The Holtzer-Cabot centrifugal shunt. When at rest the copper granules settle around the end of the spindle. When it is revolved they are thrown out by centrifugal force, thus breaking the shunt circuit.

winding coil is attached to the shaft by the pin, *C*, and the other to the pin, *D*, which is insulated from the metal of the shaft, as shown by the shaded parts around it, and leads the current to the terminal connection through an insulated path. This pin has a platinum head, which, when the apparatus is at rest, is in contact with the bob, *E*, carried on the end of the light leaf spring, *F*, which, in turn, is secured to the shaft of the shuttle by the screw, *G*. So soon as the shuttle is revolved by turning the crank, the end of the spring, *F*, naturally flies outward, impelled by centrifugal force, until the bob, *E*, comes into contact with the stop, *H*. Thus the shunt circuit is broken, and the coil of the armature is thrown into action. It is necessary that the shunt be broken; otherwise the generator would be short-circuited through the spring, *F*, and no current could emerge to ring the bells.

The Holtzer-Cabot Shunt.—Like all appliances for telephones, the automatic shunting devices are manifold in number.
One of the simplest and most effective of the more recent contrivances is shown in Fig. 105. It is manufactured by the
Holtzer-Cabot Electric Co., of Boston, Mass. On the end of
the armature shaft is mounted a small cylindrical brass box, in
metallic contact with the spindle, and partly filled with copper
granules, or short sections of copper wire, which are silver-

FIG. 106.—Magneto-generator supplied with a granule centrifugal shunt.

plated to prevent corrosion, and insure more perfect electrical
contact. So long as the armature is at rest, a circuit of the
smallest possible resistance is established between the end of the
spindle shank and the sides of the box. The slightest movement of the armature disturbs this; and, as soon as it begins to
revolve, the metallic granules are thrown outward, by centrifugal force, thus effectually breaking the shunt. Fig. 106 shows
a generator furnished with a centrifugal shunt of this descrip-

CHAPTER TEN.

THE SWITCHBOARD AND THE APPLIANCES OF THE CENTRAL STATION.

Telephone Systems, Large and Small.—As we have seen, each telephone apparatus includes a transmitting and a receiving instrument—the one to talk into, the other to receive the messages from some other telephone apparatus. It follows, therefore, that between any two apparatus there must be a line of wire suitable to convey the electric current bearing spoken messages. In a small system, with but few instruments, as in a country town or in a large manufactory, there may be a number of wires leading from each instrument to every other, so that one telephonist may call up any other in the system by simply manipulating a switch attached to his own apparatus. Such small systems are known as "party lines" and inter-communicating systems, and their operation and appliances will be fully explained in another section. At present we are concerned only with the most familiar method of connecting the talking circuits of telephones. It is known as the "central station," or exchange, system.

Grounded and Metallic Circuits.—It is hardly necessary to explain to any one that if an inter-communicating system were adopted in connection with any number of telephone stations above ten or twenty, or in any town of size and business activity, the amount of wiring necessary to complete the circuits would be beyond the possibilities of commercial expenditure. It is positively essential that, in all but the smallest systems, each apparatus have but one circuit—an incoming and an out-going wire, or an incoming wire and a ground return. The "ground return" is the method made familiar by its adoption in telegraphy. Here, as we know, the wire carrying the message current is strung on poles, or buried in cables, and the return current, that completes the circuit, flows through short lengths of wire to the ground, and thence back to the sending station.   The attachment in the ground is made either by water mains or sewers, or by sheets of metal buried at the required depth.   This system is impracticable in cities, where the ground is filled with pipe lines and other obstacles that would immensely weaken any current, or subject it to outside interferences—sneak currents and contact with other circuits—and most often destroy its power to transmit articulate sounds.   To reduce these interferences to a minimum, metallic circuits, consisting of two distinct lines of wire, of the same size and material, are most commonly adopted.   All lines are then carried to a central station, where are installed devices suitable for connecting, as desired, any two subscribers.

The Switchboard : Its Construction and Operation. — For the purpose of making these connections a device known as a switchboard is employed.   Fig. 107 gives an idea of the general appearance and construction of a common form of this apparatus.   As may be seen, it consists of two distinct parts: a series of upright panels carrying drop shutters and round apertures under the annunciator numbers; and a horizontal board or table, upon which appear a number of upright instruments, and in front of them a row of short levers.   The panel apparatus thus consists of a number of small instruments such as are used on a hotel annunciator, which are known as "drops," and of another series of instruments, fixed behind the round holes on the front of the panel, which are called "jacks," or spring jacks.   The upright instruments, that stand in rows below the panel-boards, are the " plugs "; and each of them is secured at the end of a flexible cord, passing through a hole in the switchboard table, attached at the back of the board, as shown in the next figure, and held in place by pulley weights. The object of these plugs is to close the circuits between any two subscribers' lines, one of a given pair being inserted in the.

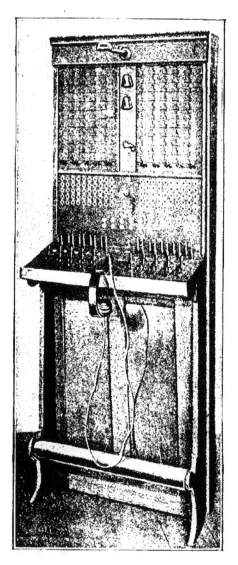

FIG. 107.—Sterling Bell-type Switchboard of 100 Drops.

"jack" corresponding to the calling station, and another in that corresponding to the called station, connection being made between the two lines by means of the flexible conducting cords. The row of small levers at the front of the table are the operator's "listening and ringing keys."

FIG. 108.—Rear view of Sterling 100-drop Bell-type Switchboard.

Making a Telephonic Connection.—The method of operating the switchboard is as follows: When any subscriber desires to have communication with any other, he will, as we have seen, operate the magneto of his instrument, thus sending a current along the line wire, which causes the shutter of the drop to which his wire is connected to fall, thus giving his number to the operator, who sits in front of the switchboard table. Since each subscriber's circuit is completed at the drop of the board, any one can understand the method of causing the shutter to drop by thinking of the operation of the ordinary burglar alarm, or of a hotel teleseme. As soon as the drop shutter falls, the operator lifts the inner, or "answering," plug of the two immediately under the row of drops in which this particular one happens to be, and inserts it in the hole of the jack bearing the corresponding number. Then, by moving the listening key in the row immediately in front of the drop and plug in question, she throws her own telephone set into circuit, and is thus able to communicate with the subscriber calling. Having learned the number of the subscriber with whom he wishes to speak, she takes the forward plug of the two immediately below the drop of the calling subscriber, and inserts it in the jack of the number corresponding to that of the subscriber called. This done, she moves the ringing key in the first subscriber's row, so as to sound the call bell of the one called; the current for this purpose being supplied, either by a hand magneto-generator attached to her section of the switchboard, as shown in Fig. 108, or else by thus throwing into circuit the wire leading to and from a power-driven dynamo attached to the exchange. The latter is the plan adopted in all large exchanges.

The Clearing-out Drop.—As soon as the operator has attracted the attention of the called subscriber, she again shifts the key, thus throwing the two into one circuit, and enabling them to have their conversation. As soon as this is finished each hangs up his receiver and turns the crank of the magneto of his apparatus, thus causing to fall another drop, usually placed at the base of the panels, which is known as the "clearing-out" drop. On receiving this signal, the operator restores the keys to their first position and removes the plugs from the jacks, allowing them to be drawn down by weights attached to their cords, to their place in front of the panel, as shown in the figure,

The Circuits of a Switchboard.—As one might easily
guess, the switchboard, like the telephone apparatus already
described, is a combination of several different circuits, each of
which works only when the others are cut out of line.   When

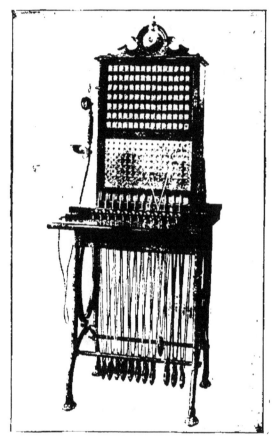

FIG. 109.—Ericsson 100-Drop Table Switchboard.

the telephone apparatus is at rest the call bells are in circuit,
so, in the switchboard, the drop is normally ready to respond
to the impulses sent along the line when the magneto generator
is set in motion. Just as the removal of the receiver from the
switch hook makes the talking circuit of the telephone appa-

ratus, so the insertion of the plug in the switchboard jack cuts out the drop circuit and makes the talking connections.

Grounded Line Switchboards.—So far as concerns the construction of the line, jack and plug, there are two kinds of switchboards: those for grounded circuits and those for metallic circuits. Fig. 110 shows the details of a grounded circuit switchboard drop and jack. Here the current from the generator of the subscriber's apparatus enters the switchboard apparatus at *a* ; thence through the leaf spring, *b*, of the jack belonging to that particular subscriber; through the contact screw, *c*, which is insulated from the metal of the jack, as shown; through the wire leading from it to the coil of the electro-magnet, *d*. The result is that the magnet attracts the armature, *e*, raising the attached lever, *f*, and thus freeing the hinged drop shutter, *g*, which falls, attracting the attention of the operator, and disclosing the number of the calling subscriber. The condition of the drop apparatus, before it is affected by the current, is shown in Figs. 116-117. Here we see that the armature, *e*, of the magnet, not being attracted, holds the position shown by the weight of the bar, *f*, which, by the hook at its end, retains the shutter, *g*, in its normal position.

Circuits of a Grounded Line.—Immediately on noticing the drop number the operator thus closes the shutter and inserts the plug in the jack. By this act, as may be understood, the metallic point of the plug forces the spring out of contact with the point, thus cutting out the drop from line, and making a new circuit through the conducting cord attached to the plug, as shown in the figure. The current, then, no longer goes to earth by the ground wire, but through the grounded connections of the two communicating telephones, when their lines have been joined by the two plugs and cords standing in the row just beneath the drop of the calling subscriber. It is essential that the two plugs, used to make connection between any two subscribers, should belong to the same pair; since each pair is connected through the conducting cords with one set of

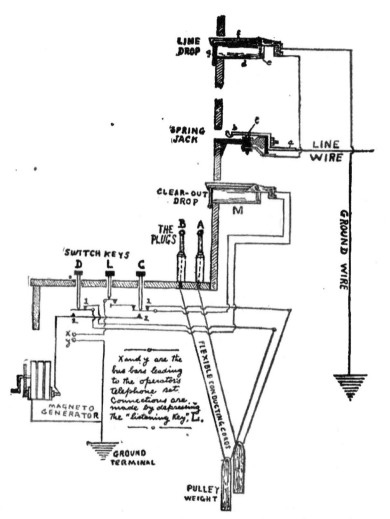

FIG. 110.—Circuits of a grounded or common-return switchboard.   The line passes through the clear-out drop, which is low-wound.   In common-return circuits, described on pages 313-314, the common-return wire is connected where grounds are shown in this figure.

listening and ringing keys, which play the parts already described.   To attempt a connection with any two plugs not in the same pair would mean failure to make a circuit between the two subscribers' apparatus.

Switchboard Plugs.—The plugs used in switchboards having a grounded circuit consist simply of pins of metal properly shaped, and joined on to the conducting cord by suitable screw connections. In a switchboard of a metallic circuit the plugs are made with double connections, in order to maintain the lines of the two wires, the line and the return, throughout. Their construction will be explained later.

Grounded Switchboard: Operator's Circuits.—The circuits of a grounded line switchboard are also indicated in diagram in Fig. 110. Each of the plugs, A and B, is connected by its flexible conducting cord with the apparatus of the ringing keys, C and D, respectively. As will be seen from this diagram, which uses the simplest forms of listening and ringing devices, each of these keys has two contacts, 1 and 2, so as to enable it to stand in line with either of two circuits. Upon perceiving that a drop bearing a number, say, 10, has fallen, the operator inserts plug A in the same numbered jack, and at the same time pushes down upon the listening key, L, so as to throw in her receiver and transmitter, R and T, thus enabling her to converse direct with subscriber number 10, the circuit being established from the ground connection of his apparatus, through the outgoing line wire, the spring jack of the switchboard, the plug, A, and its cord, to contact 1 of key C, and thence through key L, to the section talking apparatus, which is provided with receiver, transmitter, battery and induction coil, in precisely the same manner as the subscriber's apparatus, and ending the circuit in the ground connection.

Grounded Switchboard: Line Circuits.—On ascertaining that 10 desires to converse with 84, for example, she inserts plug B in jack 84, at the same time pressing key D to its contact, 2, thus throwing into circuit the magneto-generator, G, which is operated either by hand or on a "bus wire" from a power driven dynamo in the exchange. A circuit is thus made from ground in the exchange, through the generator, G, contact 2 of key D, cord and plug B, jack 84, line wire 84, apparatus

84, to ground; thence back again, causing the bell of 84 to ring. This done, the operator restores key D to its normal position in connection with its contact, 1, thus throwing 10 and 84 into a circuit which is bridged across by the clearing-out drop.   This

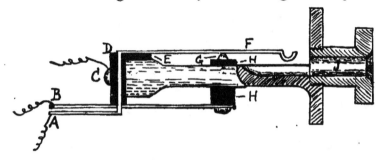

FIG. 111.—Metallic Circuit Switchboard Jack, showing double electrical contacts.   Tip contact at spring, F; sleeve contact in thimble, J; connection to line drop through anvil, G, and wire, B.

is constructed precisely like an ordinary line drop, although generally of a higher resistance—between 500 and 1,000 ohms. This high resistance and self-induction is used in order that these drops, permanently bridged across the circuit, may not shunt the telephonic current.   They are also enclosed in soft

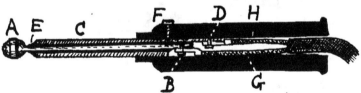

FIG. 112.—Metallic Circuit Switchboard Plug, showing double contacts. A, tip contact; C, sleeve contact : E, insulating bushing.

iron tubes, for the purpose of increasing the electro-magnetic effect, and also to prevent induction from other drops and lines.

After the two subscribers have been connected in the manner indicated, the operator may keep her listening key depressed in order to find out whether connection has been made, if 84 has answered the call.   She then breaks circuit with her own telephone set, leaving all the keys in the raised position.   So n as the conversation is ended the subscribers hang their receivers on their switch hooks, and turn the cranks of their magneto-generators, thus sending along the line a current of sufficient strength to operate the clearing-out drop, the talking current being too weak for that purpose. As soon as the

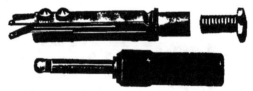

FIGS. 113 and 114.—Spring Jack and Short Answering Plug of a Sterling Metallic Circuit Switchboard. The Jack is a Brass Tube, with Springs of German Silver.

operator sees the shutter fall, she knows that the conversation is ended, and removes the plugs from the jacks.

Metallic Circuit Switchboards: Plug and Jack.—The apparatus of a metallic circuit switchboard is arranged to accomplish the same results, although differing in numerous

FIG. 115.—Metallic Circuit Jack and Plug of the Keystone Electric Telephone Co.

details of construction. As, however, metallic circuits are almost universal in present-day telephone practice, it will be necessary to examine such apparatus in detail.

Fig. 111 represents the construction of a metallic switchboard jack. As will be seen at once, its principal point of difference is that it has three points of circuit connection—*A*, *B*, and *C*. This is a feature common to all its various forms. As regards the connections of *A* and *B*, it will be readily seen that they are the same as those in a grounded circuit jack. The third wire is connected to the binding screw at *C*, and separated from the other two by the insulating blocks of hard rubber, *D*

and *E*. When the telephone circuits are not in use, the leaf
spring, *F*, connected to the terminal, *A*, is in contact with the
point, *G*, insulated from the tube of the drop by the hard rubber piece, *H*, and in electrical connection with the terminal

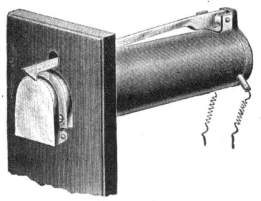

FIG. 116.—A typical Switchboard Line drop; the Tubular Drop of the Keystone Co.
The iron tube contains an electro-magnet, to whose coil the current is admitted by the
wires, as shown. The armature is hinged at the rear of the tube, and when it is attracted
raises the longitudinal lever, thus releasing the shutter.

FIG. 117.—Tubular Drop of the Couch & Seeley Co. The tube is drilled from a solid
rod of iron, and by the attachment of the magnet core gives a double-pole effect at the
armature. Connections of the night bell circuit are shown beneath the hinge of the
shutter.

wire, *B*, so that a current coming from the generator of a subscriber's apparatus through the wire, *A*, passes entirely around
the jack into the coil of the drop, as in the form of switchboard
just described. The plug used in a metallic circuit board is
similarly compound, having two metallic contacts, insulated
from one another, instead of the one used in the grounded circuit board. Fig. 112 shows the construction of a plug of this
description. Here we have the ball-pointed tip, *A*, as in the
other type of plug, for the purpose of engaging the leaf spring

of the jack, and cutting out the circuit of the drop coil. *A* is enclosed in a hard rubber tube, *E*, which acts as a bushing to insulate it from the other contact metal part, the sleeve, *C*, which, in turn, is enclosed in the rubber handle, *F*. The contact, *A*, is attached to one wire, *G*, of the conducting cord by a screw at *B*, and the other contact, *C*, is similarly attached to the second cord wire, *H*, at *D*.

Circuits of a Metallic Switchboard.—When the double contact plug is inserted in the jack of a metallic circuit switchboard, it closes a circuit having one terminal at *F*, in connec-

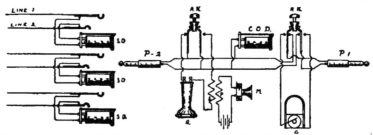

FIG. 118.—Diagram of Circuits of a Metallic Switchboard (Fig. 104).   *S. D.*, Subscribers' Drops ; *C. O. D.*, Clearing-out Drop ; Key at right of *C. O. D.*, Ringing Key ; *A. K.*, Listening Key ; *P. 1*, Calling Plug ; *P. 2*, Answering Plug ; *R*, Receiver ; *M*, Transmitter ; *P*, Primary of operator's induction coil ; *S.*, Secondary ; *G*, Generator.

tion with the wire, *A*, in Fig. 111, and the other in the tube of the jack, at *J*, through the wire, *C*, held in metallic contact with the jack by the screw, as is shown. Thus the speech-bearing current enters at *A*, and returns to line at *C*, the current being continued through the two wires of the plug cord, and completed in the operator's table apparatus, her talking set, or the line and apparatus of another subscriber, in a manner similar to the system previously described.

Metallic Switchboard : Operator's Circuits.—The circuits of a metallic circuit switchboard are shown in Figs. 118-119. Here, as in the former figure, the inner is the "answering plug," to be inserted in the jack of the calling subscriber, and the outer the "calling plug," which is intended to be inserted in the jack of the subscriber who is called for. As

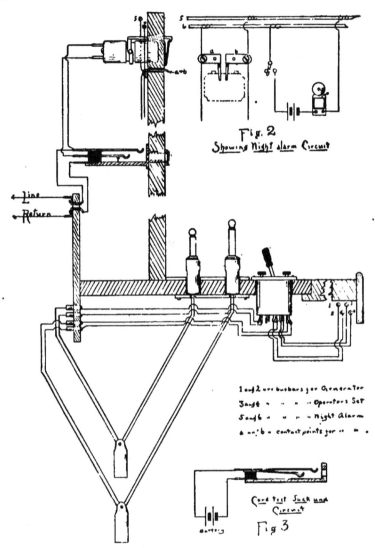

FIG. 119.—Circuits of a Metallic Switchboard, showing the mechanisms for operating the talking and calling connections. This figure illustrates the wiring of a Sterling Metallic Switchboard, which will be explained later.

may be seen, there are two wires from each plug, corresponding to the double contacts, tip and sleeve, of each plug. They may be traced through the flexible cords, which are held down by the pulley weights, when not in use, and terminate in the springs projecting from the bottom of the combined "listening and ringing key," as shown in Fig. 119. The construction of this and other switchboard keys will be explained later, but it is intended to accomplish the self-same results as are accomplished by the three keys, *C, D* and *L,* in Fig. 110 of the grounded circuit board.

Operation of the Switch Key.—By the use of the lever of this key, the circuits of the two plugs, or of either of them, may be changed, as in the other type of board, and, by the use of the two press keys, on either side of the lever, the ringing generator may be thrown into the circuits in precisely similar fashion. These results are accomplished, as may be readily understood, by the fact that the circuits of the ringing generator, of the operator's telephone set and of the clearing-out drop, also terminate in the key, as indicated. This mechanism enables the same program to be followed as in the grounded circuit board, after the answering plug has been inserted in the jack of the calling subscriber, to wit: 1. The cutting-out of the drop; 2. The throwing-in of the operator's talking set; 3. The insertion of the calling plug in the jack of the called subscriber; 4. The making of the ringing circuit with the apparatus of the called subscriber; 5. The connection of the two lines with the clearing-out drop bridged on their circuit.

The difference to be constantly borne in mind, in comparing the metallic with the grounded circuit switchboard, is that the latter must be so wired as to permit of all circuits made ending in a grounded connection, either at the exchange or at the subscriber's apparatus; the former is so constructed that all circuits, short or long, are composed of two metallic lines throughout, permitting every current used to emerge from and return to its source direct.

CHAPTER ELEVEN.

THE OPERATOR'S SWITCH KEYS AND TELEPHONE SET.

Combined Listening and Ringing Keys.—Since, in order to meet the needs of every calling subscriber, as we have seen, the operator must perform no less than seven distinct acts in the way of shifting and changing circuits, it has been the constant effort of inventors and manufacturers to produce devices to simplify her work, by allowing the greatest number of results to be accomplished with the fewest movements of her hands and arms. The object is, not only to save the operator's strength, but also to economize time in the "rush hours" at an exchange. Thus it is that a great variety of keys has been devised, on almost as many different principles; and attachments for making the operator's listening and ringing circuits have also been combined with the plugs and jacks. It has been found, however, that an experienced operator can do the necessary work with the ordinary devices, quite as readily and easily as with some others intended to save her work.

The Cook Switch Key.—The key shown in Fig. 120 is of the form known as the Cook key, invented by Frank B. Cook, and manufactured by the Sterling Electric Co., of Chicago. The interior view of this kind of key is shown in Fig. 121, which also shows its details and operation. Its mechanism consists of five pairs of metallic spring contacts, ·a hard rubber insulating partition running in the length of the case, serving to divide it into duplicate halves—that is to say, one of each pair of springs is set on either side of the partition. The arrangement may be understood from the exterior view of the key.

Its Construction.—Turning, now, to consider the details shown in Fig. 121, we find that the lever, *A*, ends in a cam, *B*, turning on its pivot, and so shaped as to work on the two springs, *C* and *D*, as shown. This cam extends through the hard rubber partition, so that the changes effected on one side

E C G D F

FIG. 120.—**Cook Listening and Ringing Key.** A hard rubber partition runs in the length of the case, so that it divides the instrument into two equal parts; one *E*, one *C*, one *G*, etc., being on either side, making double connections, as shown.

are also effected on the other; thus, in studying its operations, we must bear in mind that there are two springs *C*, two *D*, two *E*, two *F*, and two *G*, each pair, one spring being on each side of the partition, forming the terminals of some particular circuit. The pair of springs, *E*, terminate in the tip and sleeve of the answering plug; the pair *F*, in the tip and sleeve of the calling plug; the pair *C*, in the coil of the clearing-out drop; the pair *G*, in the magneto-generator. The operator's desk telephone set is connected with the springs, *D D*, opposite to *C C*. The pair *E*, and the pair *F*, being always in contact with the pair of double springs, *H*, which are connected together, as shown by the upper dotted line in the diagrams, and make a permanent circuit between the tips and sleeves of the two switchboard

plugs. The lever being pressed all the way to the left, the
coil of the clearing-out drop is bridged into the plug circuit,
as is the case when two subscribers are conversing.' Thus at
the completion of any conversation the key mechanism is in
position for another call without alteration.

Its Operation.—When a subscriber calls, the operator
inserts the answering plug connected with the pair of springs,

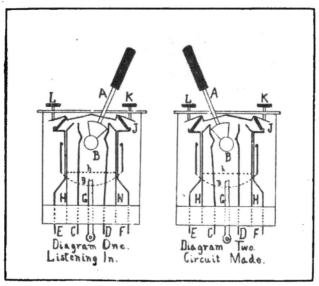

FIG. 121.—Diagram of interior construction and operation of the Cook Key showing
the relative positions of the cam and springs on the (operator's) left side of the insulating
partition within the key case. The continuity of the pair of springs, *H H*, is indicated
by *h*. Continuity of *G* and two outer springs, by *g*. The pair, *E*, and the pair, *F*, on
either side of the partition, are always in electrical contact with the double pair, *H H*.

E, into the jack, and by moving the lever into the position
shown in the first diagram of Fig. 121, bridges in her speaking
set, so as to make a circuit with the apparatus of the calling
subscriber, through the jack, from the tip and back again to
the sleeve of the plug. Having learned the number of the subscriber desired, she inserts the plug connected with springs, *F*,
in his jack, and presses key, *K*, thus forcing the pair, *F*, out of
their normal contact with the double pair, *H*, and into contact
with the pair, *G*, thus throwing in the ringing generator, which

sends a current along the line to his apparatus, through the conducting cord, plug, jack, line wire, his ringing apparatus, and back to the switchboard generator. This done, she releases the key, *K*, thus restoring the springs, *F*, to their normal contact with the springs, *H*, and breaking the generator circuit; at the same time bringing the lever, *A*, to the opposite position, as

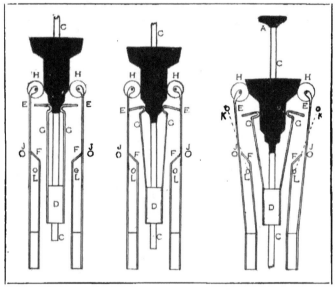

FIG. 122.—The O'Connell Switch Key. First diagram, normal and talking; second diagram, listening in; third diagram, ringing up the called subscriber. By compressing the plunger to make contact of EF and KK, both subscribers are called.

shown in the second diagram of Fig. 121, thus making a circuit between the apparatus of the two subscribers, and bridging in the clearing-out drop, connected to the pair, *C*.

As soon as she has ascertained that connection has been established, she moves the lever to the position of diagram one, its normal position, thus cutting out her set, and leaving the two subscribers to continue their conversation. If, at any time, she wishes to "listen in," in order to find whether they are done talking, she can do so by bringing the lever to the opposite position, as shown in the first diagram of Fig. 121. Whenever she has occasion to call up the calling subscriber, as, for example, when there is a delay in making connection with the subscriber called, both plugs being in the jacks, she can do so by depressing

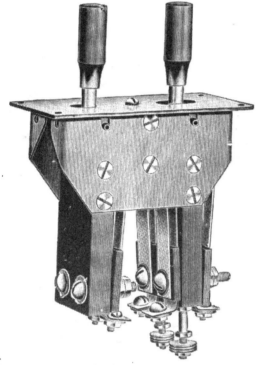

FIG. 123.—Combined Listening and Ringing Key of the Keystone Electric Telephone Co.

the key, L, thus throwing the pair of springs, E, out of contact with H, into contact with G, and sending a ringing current to his apparatus. The conversation finished, and the shutter of the clearing-out drop having fallen, she has only to close the latter, and remove the plugs, leaving the lever in the position in which it was found. By the use of this key, all the operations, except the insertion and removal of the plugs, can be performed by the operator, with her hand resting on the lever of the key.

The O'Connell Switch Key.—Another form of switch-key, which has been in extensive use on metallic circuit switchboards, is the O'Connell key, shown in Fig. 122. Its advantage lies in the fact that all necessary shiftings of connection may be performed by depressing or raising a single push key, *A*, which operates the suitably shaped wedge of hard rubber mounted on the rod, *C*, which in turn, moves through slots in the top of

Fig. 124.—Combined Listening and Ringing Key of the Western Telephone Construction Company.

the switchboard table, and in the piece, *D*. On either side of this wedge, and in such positions as to be engaged by it, are three pairs of springs, *EE*, *FF* and *GG*. The pair, *FF*, are connected with the tip and sleeve of the answering plug; the pair, *EE*, with the tip and sleeve of the calling plug; the pair, *GG*, attached to *D*, close the circuit of the operator's talking and listening set. The two pins, *JJ*, placed outside of *E* and *E*, are the bus bars of the switchboard magneto-generator. The springs, *EE*, bear the rollers, *HH*, in order to facilitate the movement of the wedge on the springs.

Its Operation.—The normal, or resting, position of the instrument is shown in the first figure. Here we find the springs, *FF*, in contact with the springs, *EE*, thus leaving all connections ready between the terminals in the two plugs, through the coil of the clearing-out drop. The two springs, *GG*, rest against the smallest portion of the wedge, as shown, thus leaving open the circuit of the operator's telephone set.

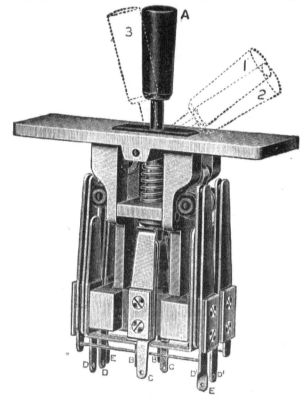

FIG. 125.—Combined Listening and Ringing Key of the Couch & Seeley Co.

As soon as a call has been received, and the answering plug has been inserted in the jack of the calling subscriber, the operator presses the key so that the wedge is forced down sufficiently to permit the next larger contact to engage the two springs, *GG*, thus throwing in her telephone set, for the purpose of learning the number desired by the calling subscriber. This done, she inserts the answering plug in the jack of the called number, and, in order to call up, again presses the key until the springs, *GG*, ride upon the third contact surface of the wedge, and the rollers, *HH*, on the springs, *EE*, move upon its widest portion. The latter action forces the springs, *EE*, into contact with the bus bars, *JJ*, of the generator, breaking the normal contact between *EE* and *FF*. Consequently the ringing current is sent through the calling plug to the apparatus of the called subscriber, the circuit of the calling subscriber being meantime cut out. If, at any time, there is a delay on the part of the called subscriber, and it is desirable to call the calling subscriber, when connection is made, both at once may be called by pressing the key downward until the rollers of *EE* ride on the extreme edges of the wedge, *B*. This act forces the springs, *EE*, still further back, and brings them into contact with the pins, *KK*, which are in insulated connection with the two other pins, *LL*, against which the springs, *FF*, rest, after the pressure of *EE* has been withdrawn by the downward movement of the wedge, *B*. Thus the generator current flows from *J*, through *E*, to *K*, thence to *L* and *F*, through the cords, plugs, jacks, lines and ringing bells of both subscribers, and back again, on the other side, to the generator.

As soon as connection has been established, the operator depresses the key to the position shown in the second diagram, thus establishing a talking circuit between the two subscribers, while still keeping in her own telephone set, as shown, so as to listen in. Having ascertained that all is right, she moves the key again to the position it held before the call came, thus leaving the line free for conversation without the bridging-in of her own set. This is the final position of the key.

The Couch and Seeley Key.—Fig. 125 illustrates the self-restoring, combined listening and ringing key recently introduced by the Couch & Seeley Co., of Boston. Unlike the keys just described, the operating lever, *A*, is automatically restored by the force of the spiral spring shown in the cut. Like the Cook key, however, it has five pairs of contact springs, which

FIG. 126.—Attaching rod for suspending a switchboard operator's transmitter. It is secured to the top of the switchboard, as shown in Fig. 107, by the screw plate, the flexible conducting cords passing through the pulleys and being held back by pulley weights.

FIG. 127.—Operator's Headgear and Receiver.

FIG. 128.—Ericsson's Breastplate Transmitter for a Switchboard Operator.

form the terminal connections. Thus C and C connect to the operator's telephone set ; B and B, directly back of them, to the clearing out drop ; E and E, to the generator ; D and D to the answering plug; D^1 and D^1, to the calling plug. The operation is as follows: On receiving a drop-call, the operator inserts the answering plug in the corresponding jack, at the same time throwing the lever to position 1, thereby switching in her telephene set. Having learned the number of the subscriber called for, she inserts the calling plug in his jack, at the same time depressing the key to position 2, thereby throwing the generator into line and cutting out her talking set. The lever restores itself automatically from position 2 to position 1, and only a slight additional touch is required to restore it to the normal position, A.

Other Switch Keys.—The ringing and listening key of the Keystone Electric Telephone Co. operates on the same principle of positive contacts, but, as shown by Fig. 123, it has two levers instead of one. These are, however, so constructed as to restore automatically or maintain the circuit, as desired.

Still another highly efficient type of key, that of the Western Telephone Construction Co., is shown in Fig. 124. With it the listening connections are made by a lever, and the ringing connections by the push button. In this form of key the cam operates on the principle of sliding contacts.

Switchboard Operator's Telephone Set.—The talking set of a switchboard operator consists of a receiver of the watch-case pattern, previously described, which is secured to her left ear by the head band attachment, as is shown in Fig. 127, and of a transmitter suspended in a convenient position before her from the top of the switchboard cabinet. The attachment of this apparatus is shown in Fig. 126. Another form for attaching the transmitter, which is much used in Europe, is shown in Fig. 128. Here the transmitter battery circuit is made, and ready for talking when in the position shown; it is broken by the operator turning the transmitter on its hinge until the

mouthpiece is against her breast. This form has the advantage of keeping the transmitter always in a convenient position for immediate use, saving the operator some inconvenient

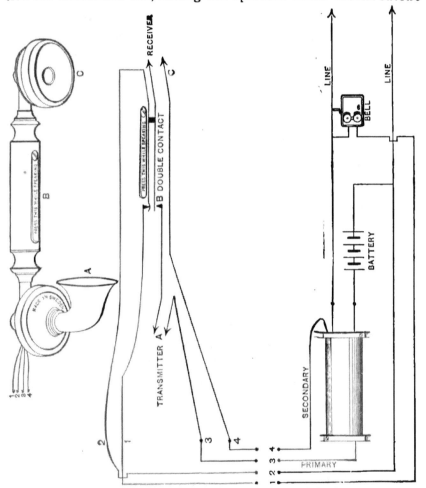

FIGS. 129 130.—Ericsson Hand Microtelephone and circuits.

movements of the hand to hold the transmitter steady, as is
frequently necessary when it is suspended on a cord before her.

The Microtelephone.—Some switchboard exchanges having only about 100 or 200 subscribers, which is a number not

likely to keep an operator constantly busy, are provided with an ordinary "microtelephone" set, such as is shown in connection with the switch board in Fig. 109. This form of instrument, which has been in extensive use in Europe for a number of years, and is now gaining favor in this country, is shown in Fig. 129. The circuit connections are displayed in Fig. 130. The battery circuit of the transmitter is closed by pressing the key on the handle,

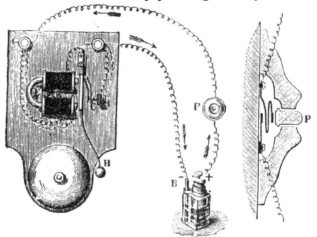

FIG. 131.—Direct current electro-magnetic call bell of the type used on switchboard night alarm circuits. The figure shows the difference between this and the polarized ringer of the telephone apparatus, as already described. The circuit is closed by making contact of the two terminals, after the manner of the push button here shown in section. The current then energizes the electro-magnet, causing it to attract its armature, but before the hammer has sounded the gong the circuit is broken at the spring, c, the momentum of the hammer then carrying it forward the rest of the way. After sounding, it again springs back, thus making the circuit anew and being again attracted by the magnet.

labeled, "Press this while speaking." Thus, as may be readily seen, the spring, *B*, is forced into contact with the anvil of the transmitter wire below it, and the talking contacts are complete. This particular form of microtelephone is manufactured by the Ericsson Telephone Co., which claims priority in the design, and was the concern that introduced it in Europe. Microtelephones are frequently arranged for "desk sets" in business offices, in which case the annunciator call bells are hung upon the wall above, the attachments being made as indicated in the figure of the circuits,

The Switchboard Night Bell.—In small exchanges at most times, and in large exchanges at night, during dull hours and on Sundays, it frequently happens that the operator is absent from her seat before the switchboard, and, hence, unable to note the fall of a subscriber's drop. To remedy this difficulty a call bell circuit is attached to each switchboard section, the gong being usually placed at the top of the panels, as indicated in Fig. 109. The wires are so arranged that the ringing circuit may be made by the shutter of each drop falling open against two pins arranged directly beneath, as shown in Fig. 119, thus closing the circuit through the metal shutter. During the day, or in

FIG. 132.—Ordinary two-point lever switch of the type used in connection with switchboard night alarm circuits, as shown in Fig. 107.

busy hours, the circuit connections of the night alarm bell are broken by a switch of the type shown in Fig. 132. This renders the bell inoperative even when the drop shutter rests against the contact pins.

Switchboards for Small Exchanges.—In a very small exchange, of 100 or 200 subscribers, the services of one operator are usually sufficient. Thus ordinary switchboards, such as are shown in the figures, are made with about 100 drops. In stations of 5,000 subscribers, or more, where there is a corresponding volume of business per subscriber, a number of such operators' "positions," as they are called, of 100 each, are arranged in a long row, each operator attending to the requirements of 100 subscribers.

Large Exchanges: Transferring Calls.—If, in such an extensive exchange, one subscriber desires to communicate with another wired to a different position, under the management of a different operator, there are two ways in which the result may be accomplished; either by the use of the device known as the " multiple switchboard," which will be presently described, or by some system of inter-communicating transfer. By one system of transfer each position is provided, not only with subscribers' drops and jacks, but also with drops and jacks corresponding to the other positions of the exchange. When a subscriber, say, in position 1, desires to communicate with one who has his drop in position 5, the operator at position 1 calls the operator at 5, and makes the talking circuit between the two subscribers by means of a combination of devices which will be fully explained in the proper place.

Trunking Connection.—For the purpose of " trunking out," or connecting with another exchange, a special section is provided in all switchboards, which is in communication with every other section in the same exchange. The system of inter-communication just mentioned is, in general, fairly descriptive of the method of connecting the subscriber in one exchange with the subscriber in any other. In the telephone systems of large cities, each exchange has a number of connections with every other, in order to accommodate the vast trunking business.

CHAPTER TWELVE.

IMPROVED SWITCHBOARD ATTACHMENTS.

Labor-Saving Devices.—As the result of constant efforts to simplify the operations necessarily performed by the switch-board operator in making the circuits desired by subscribers, and attending to other business connected with the switching apparatus, a number of improved devices have been introduced by the various manufacturers of telephones and supplies. Among such may be mentioned subscribers' drops so arranged as to act as clearing-out drops as well, self-restoring call-drops, and combined drops and jacks. The advantage in every case is that the operator is saved a considerable expenditure of energy and time in having fewer points to observe in the course of her work. Thus, in the combined jack and drop, she has to insert the plug in an orifice indicated by the falling of the drop shutter, instead of searching for the jack bearing the same number. This is a desirable saving of time and nerve force. The same is true in the use of combined calling and clearing-out drops. Here she is saved the necessity of paying attention to more than one series of drops at a time, knowing in an instant whether the falling of a shutter means a call or a clearing out in each par-ticular case, and avoiding much of the delay due to pressure of work in the rush hours.

Combined Calling and Clearing-out Drops.—In general, when the calling and clearing-out drops are combined in one instrument, the only changes of construction are in winding the drop coils to a higher resistance, to prevent the short circuiting of the speaking current; enclosing them in iron tubes, or cap-sules, to neutralize the influence of outside induction, and in arranging the lines between the drop and its corresponding jack on a different plan. As we have previously learned, the resistance of the coil of the ordinary clearing-out drop is 500 ohms, or over, while that of the ordinary calling drop is about 80 ohms. In order, therefore, that the requisite resistance may be inserted in the line, an arrangement has been adopted whereby the drop of the called subscriber is cut out of line, while that of the calling subscriber is left bridged in to act as a clearing-out indicator. The means adopted in the boards of the Sterling Electric Co. for accomplishing this result is to make the calling plug longer than the answering plug, in order that it may actuate the mechanism of the jack to a greater extent, and thus cut out its drop, after making the tip and sleeve connections also formed by the answering plug.

The Sterling Switchboard System. — The system of wiring adopted in the circuits of this company's switchboard is shown in Fig. 119. As may be seen, the jack is constructed with three electrical connections: the line wire, ending in the upper leaf spring; the return wire, ending in the base of the jack; and the return wire of the drop coil, ending in the second leaf spring. Insulating blocks separate the two springs from one another, and from the base of the jack. To insert the short answering plug in the jack would mean merely to make tip contact with the upper spring, leaving the return connection for the drop circuit, and thus bridging the drop across the talking circuit formed by the tip and sleeve of the plug. To insert the long calling plug in the jack would mean to raise the lower spring into contact with the upper, thus short circuiting the drop coil belonging to that subscriber, and leaving line connections only through the jack.

The Sterling Drop. — The switchboard appliances of the Sterling Electric Co. are interesting as combining a number of ingenious and effective contrivances, which deserve description. The tubular bridging drop is shown in Fig. 133. As may be understood, its construction is different from that usually employed, in that the armature is swung midway on the enclosing sheath, which is constructed in three pieces, as shown in

Fig. 134. The forward cap being secured to the magnet core, derives polarity so soon as a current is sent through the coil, and attracts the pivoted armature with the usual result of releasing the shutter. The restoring mechanism consists of a lever fixed beneath each back of ten drops, and just above the jack panel. As may be understood by reference to Figs. 133 and 135, each drop has, secured back of the falling shutter, a " restoring slide," with flange at top and bottom. Reference to Fig. 107, which shows a typical Sterling switchboard, will reveal the fact that the restoring slide of the topmost drop rests on the slide of the one beneath, so on down, and that the slide of

FIG. 133.—Sterling Line Drop, showing the centrally swung armature and the restoring slide in front.

FIG. 134.—Sheath of a Sterling Drop, showing the three parts of which it is composed.

the lowest rests on the restoring lever, any movement of which will actuate the slides of the whole series, thus raising the shutters. By the attraction of the armature the drop lever raises the slide sufficiently to release the shutter, which is held in its normal position by the upper flange. These drops are mounted in rows of five, as shown in Fig. 135.

Sterling Jacks and Plugs.—Fig. 113 shows the Sterling jack, which is made of solid brass casting, with German silver springs and hard rubber insulation. Fig. 114 shows the short answering plug used. Reference to the cut of the board will show the excellent arrangement of the rows of plugs, which are " staggered," or so disposed that each plug in the rear row is opposite the space between some two of those in the front row, thus enabling the operator to select the short answering plug without inconvenient contact with the long calling plugs in front.

The Self-Restoring Drop.—The self-restoring switch-board drop consists of two electro-magnets, set end to end, so that there is a magnetic action at both front and back of the instrument. The first magnet is actuated by the generator current from the calling subscriber, and is in all respects like the ordinary drop magnet, previously described, having a pivoted armature, attached to a rod extending to the front of the instrument, where a hook engages a drop shutter. Attraction of the hinged armature by the magnet causes the attached bar to rise, thus releasing the shutter from the hook, and allowing it to fall open. The variation comes from the fact that the

FIG. 135.—Strip of five Sterling Drops. This is the most usual method of mounting switchboard drops of all descriptions.

mechanism which is the drop shutter in the ordinary coil is, in the self-restoring variety, the armature of the forward magnet, and itself controls the operation of another shutter hinged at the top. The current for the second magnet coil is supplied by a local galvanic battery in a manner to be presently explained.

'ts Construction and Operation.—Fig. 136 illustrates the mechanism and operation of the self-restoring drop. Here *O* is the operating magnet, such as is arranged in the ordinary type of drip. *R* is the restoring magnet, and *B* the common base-plate to which both are secured within an iron sheath. *A* is the armature of the operating magnet, and is pivoted at *C*, where it is also attached to the bar, *D*, bearing the hook, *H*, at its opposite end. This hook rests in a groove at the top of the shutter, *Aa*, which is pivoted on its lower end at *P*. On the upper side of this same magnet is pivoted the indicator flap, *F*, in such fashion that it normally hangs directly in front of *Aa*. Now, when the generator current enters the coil of the operating magnet, *O*, it causes the armature, *A*, to be attracted, with the result that the bar, *D*, is raised, thus releasing the shutter, *Aa*, from the hold of the hook, *H*, and allowing it to fall outward on its hinge. By this movement the flap, *F*, is also forced out, and takes the position shown in Fig. 137, being held there by the weight of *Aa*. As soon as the plug is inserted in the jack corresponding to this drop coil, a current is sent through the coil of the magnet, *R*, whose core projects into a recess in the center of the armature shutter, *Aa*. By the action

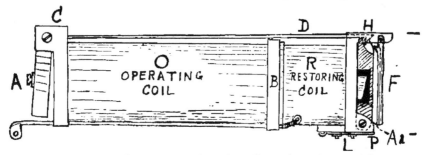

Fig. 136.—Diagram showing the mechanism of a self-restoring switchboard drop. The figure shows the normal position of the drop shutter.

of this current the shutter, *Aa*, is attracted to the pole of the magnet, *R*, taking the position shown in the first of the two figures; the hook, *H*, again engaging the notch at the top of *Aa*, and the flap, *F*, again falling to the first position by its own weight. So long as the current is continued through the coil of *R* by the presence of the plug in the jack, the shutter, *Aa*, is held fast, and any current coming to the coil of *O*, while it may attract the armature, *A*, and actuate the bar, *D*, cannot operate the drop.

Such an attachment as this must greatly increase the complexity and original expense of a switchboard, but so needful is it that the duties of the operator be simplified as much as possible, that these matters are of very small importance in comparison.

Circuits of a Self-Restoring Drop. — In the circuit arrangements used in connection with a self-restoring drop, the operating coil, *O*, is permanently bridged across the talking circuit, its high resistance and retardation serving to shunt the telephonic current onto the line, while, at the same time, it is always open to ringing currents from the hand generators in the subscriber's apparatus. The need for cutting out the coil of the drop is obviated by the fact that it is rendered inoperative, as we have seen, so long as the plug is in the jack, sending a current through the coil, *R*. Consequently the jacks are made with but the usual two contacts for the tip and sleeve of the plug, respectively. The form of plug used is the same as that

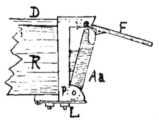

FIG. 137.—Shutter of a self-restoring switchboard drop, showing the position of the shutter after a call has been received.

previously shown, for the ordinary metallic circuit switchboard, with the exception that it carries a metal ring, or collar, outside of and insulated from the sleeve contact previously described.

This second sleeve, or insulated ring, is intended to close the battery circuit by connecting the two terminals of the restoring battery circuit, which are two thimbles at the entrance of the jack, so that, by inserting the plug in the jack, we not only make the talking circuit through the jack springs, but also allow the current from the restoring battery to flow through the wires to the inner thimble, across this outer sleeve to the outer thimble, thence to the coil of *R*, energizing the magnet and holding the shutter armature fast, as already described. The coil of *R*, being of low resistance, would greatly injure the battery, if connected direct to the wire coming from *D*, and to obviate this difficulty the needed resistance is introduced into th circuit by connecting the wire at the hinge of the drop shutter, *F*, and allowing the current to flow through the shutter, the armature, *Aa*, and thence through the hinge into the coil of the magnet. The circuit is completed through the earth and back to the restoring battery.

Combined Drops and Jacks.—Another piece of improved switchboard apparatus, having the same object—saving the operator's time and simplifying her movements—is the combined drop and jack; which is to say, the two instruments made so as to occupy the same space on the switchboard panel. The

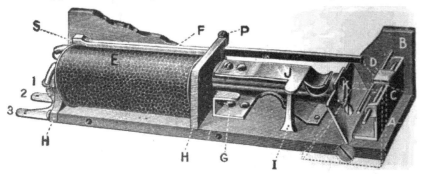

FIG. 138.—Combined Drop and Jack of the Western Telephone Construction Co.

arrangement has the added advantage of saving room, and enabling a larger number of subscribers' drops to be mounted in a given space. Several types of combined drop and jack have been placed on the market, each having its particular points of excellence and advantage.

Western Drop-Jack.—Fig. 138 represents the combined drop and jack manufactured by the Western Telephone Construction Co., of Chicago. Here the magnet, *E*, and the jack-piece, *J*, the latter a solid brass casting, are secured to the base piece, *A*, of hard rubber. The armature of the magnet is attached to the bar, *F*, pivoted at *P* to the head piece, *H*, being, however, normally held away from the pole of the magnet by a small leaf spring, *S*, bearing against its length, which effectually prevents "freezing" or "sticking" to the pole. When the instrument is not energized by a calling current, the bar, F, holds up the shutter, C, which in this cut is shown to have been thrown down by an incoming current. C is hinged to a bar running crosswise

FIG. 139.—A 300-drop switchboard of the Western Telephone Construction Co., equipped with the type of drop-jacks shown in the last figure.

to the opening between B and B, and works up or down on the curved dotted line, as shown. When a current energizes the coil, the armature is attracted to the pole of the magnet, thus causing the bar, F, to move sidewise, sliding from under the shutter, C, and allowing it to fall into the position shown in the cut.   The fall of *C* gives signal to the operator, who then starts to insert the answering plug, with its tip pushing *C* upward on its pivot into the former position supported, by *F*. The entrance to the jack is then made through the orifice, *K*, the sleeve connection being with *J*, and the tip connection with the spring shown below.   By the former connection, the spring, *J*, is lifted away from the anvil, *I*, thus cutting out the coil of the drop, as in the ordinary type of switchboard having drop and jack separate.   Although not properly a self-restoring drop in the sense in which the term is generally understood, this instrument accomplishes the same result by the simple act of inserting the plug into the jack.   It is, therefore, called a "mechanical self-restoring drop."   The bus bars of the night bell circuit run through the two side pieces, *B* and *B*, one of them serving to hinge the shutter, *C*, and the other, running parallel, so as to be engaged by a lug, *D*, carried on the top of *C*, thus closing the circuit, when the wire connections are made with the switch previously mentioned.

Fig. 139 shows a 300-drop switchboard containing this type of drops and jacks.   As will be seen, the drop shutters are represented in the raised position, thus disclosing the opening to the jack apparatus shown at *K* in the previous figure.

Connecticut Drop-Jack.—Another type of combined drop and jack, manufactured by the Connecticut Telephone and Electric Co., of Meriden, Conn., is shown in Figs. 140 and 141, the former showing the complete instrument with plug inserted in the jack springs, and the latter, the several parts.   As may be seen, the drop mechanism works in the usual way, attracting an armature, and thus actuating a lever bar with hooked end and releasing the shutter, *F*, which is normally held up by the hook and falls outward and downward on its hinge, when the hold is released.   The line wires are attached to the two terminal pieces, *I* and *I ;* the hither wire being secured to the head piece of the coil by a screw, as shown, the further wire passing in the same manner, but being insulated from the metal parts by a

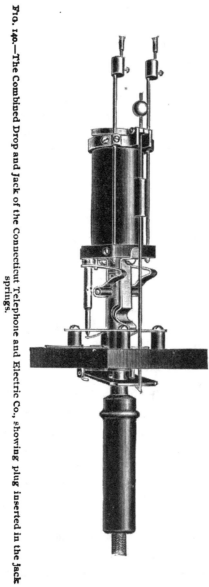

FIG. 140.—The Combined Drop and Jack of the Connecticut Telephone and Electric Co., showing plug inserted in the Jack springs.

bushing of non-conducting material. The armature is hinged, as shown, by the screw pin, *E ;* the bar carrying the hook to engage the shutter bears a rubber insulation, *A*, which serves to cut off the current of the night bell circuit from the jack and drop parts. The entire instrument is held in position on the hard rubber panel of the switchboard, *G*, by the single nut, *B*, which engages the thread on the forward end of the jack tube. The magnet coil of this drop is a separately wound helix, and may be slipped off of the core and removed whenever necessary,

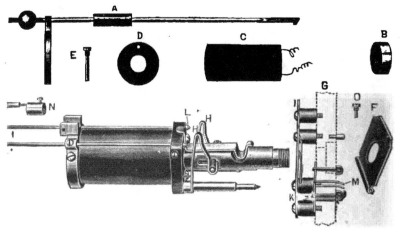

FIG. 141.—Details of the drop-jack of the Connecticut Telephone and Electric Co.

without disturbing the rest of the mechanism, the fiber disc shown at *D* sufficing to hold it in position in the tube. In action the drop coil is bipolar, the core acting as one pole and the iron shell, or sheath, as the other. The shutter is mechanically self-restoring, being restored to its normal position, in engagement with the hook at the end of the armature lever, by the simple act of raising it in order to insert the plug through the orifice into the jack springs behind. This brings it to the position shown in the first figure, which has the plug inserted and the shutter up and engaged. The tube of the jack has two contact springs, at top and bottom, as shown, to make positive tip and sleeve connections with the plug.

Operation of the Ringing Circuit.—The operation of this drop-jack shows it to be one of the most complete and ingenious switchboard contrivances on the market. The insertion of a plug between the jack springs bridges the drop coil and makes

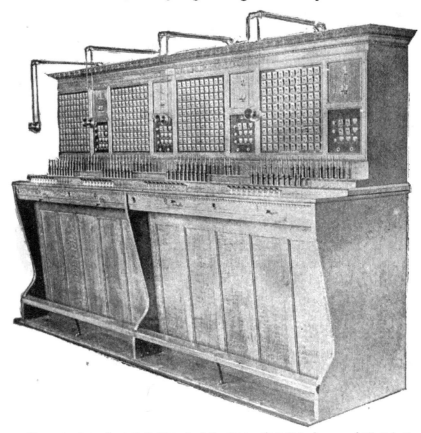

Fig. 142.—A 400-drop Switchboard of the Connecticut Telephone and Electric Co., equipped with the type of drop-jack shown in Figs. 140-141. This board has four "trunking" panels, as shown, giving a capacity for 600 drop-jacks.

the talking circuit, as in any other type of jack and drop. In order to ring up a desired subscriber, the operator has only to push the plug further into the jack, thus making contact with the cross piece attached to the springs, *H* and *H*, causing it to slide backward in its slot, and bring these springs into contact with the tips of the line wires immediately behind.   The vertical strips, *J* and *K*, are in contact with the bus bars of the gene-rator, and each of them has a tongue cut out of its length, against which bear the points of the bars carrying the springs, *H* and *H*, as shown in Fig. 140.   By this contrivance the line wires of any particular subscriber may be bridged across the generator circuit, whenever desired.   The plugs used in connec-tion with this instrument have a collar of insulating material midway on the shank, and upon this the line springs of the jack are caused to ride whenever the ringing connections are made, as described.   After giving the necessary rings the operator releases the pressure on the plug, thus allowing it to ride back into metallic contact with the springs of the jack, as it is forced outward by a spiral spring enclosed in the tube back of the cross piece attached to *H* and *H*.

Its Night Bell Connections.—The night bell connections are made by the springs, *M*, which engage an ear attached to the lower part of the shutter, *F*, thus closing the circuit.   In addition to this, the stop hook below the shutter is connected to one side of the circuit, and the eyes that hold the trunnion screw are attached to the other, thus making double contacts for the night bell.   This is an arrangement worthy of consider-ation, as giving additional assurance that the circuit will be made.

Connecticut Switchboards.—Fig. 142 shows a four hun-red-drop switchboard containing drop-jacks of the kind just described.   All shutters are in the raised position, showing orifice in each for the insertion of the plug.   The plugs are arranged in double bank, as with most up-to-date switch-boards, thus insuring greater ease in reaching and manipulating them.   The switch keys are arranged at the front of the table.   Fig. 143 shows the rear of a board of this company, ex-hibiting the method of arranging the drop-jacks.   The induction coil of the operator's transmitter is shown secured to the top of the case, and below it the gong of the night bell.

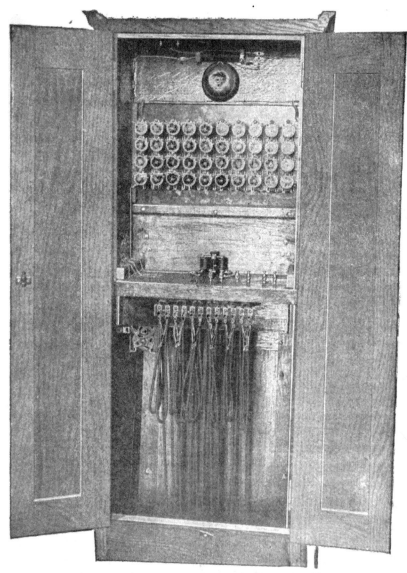

FIG. 143.—Rear view of a 50-drop switchboard of the Connecticut Telephone and Electric Co., showing forty drop-jacks installed. This cut shows the night bell gong and operator's induction coil at the top of the cabinet and the hand magneto generator below the shelf at the left.

Couch and Seeley Drop-Jack.—The recently introduced
drop-jack of the Couch & Seeley Co., of Boston, Mass., is shown
in Fig. 144. It combines the advantages of simple construction
and considerable strength. As will be readily understood, *A* is
an iron tube, to the forward end of which is screwed the core of
the magnet coil. As in the instrument just described, the open
end of the core forms one pole and the tube the other, thus

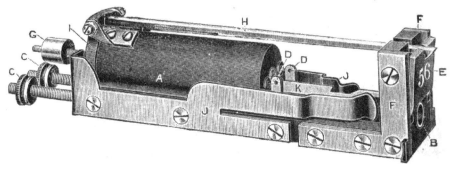

Fig. 144.—Combined Drop and Jack of the Couch & Seeley Co.

making the magnet in reality bi-polar. The ends of the windings of the magnet are brought through rubber bushings in the
tube and soldered to the tags of drop springs *D* and *D*, which
are in connection with line springs *J, J,* when the drop is in
normal condition. The line terminals are shown at *C, C.* When
a plug is inserted at *B,* into the jack frame, *K,* the line springs,
J, J, are separated from the drop springs, *D, D,* thus cutting
the drop entirely out of circuit. The shutter, *E,* has two spurs,
one on each side, formed at right angles with its face, which,
when it falls, come in contact with night alarm springs fastened
to the side pieces, *F, F.* These side pieces are of brass, and
carry the night alarm circuit to upright brass bars so notched
out as to receive the drops when mounted up in regular form in
a complete switchboard. The wiring of the complete board
using this drop is, therefore, practically done when the drops
are set in place. The armature, *I,* is provided with the counter-
weight, *G,* by which a very close and sensitive adjustment can

be made. It will throw the shutter on a bridging line with a 25-ohm shunt, the current being furnished by an ordinary 3-bar generator.

Phœnix Drop-Jack.—Fig. 145 shows the drop-jack manufactured by the Phœnix Electric Telephone Co., of New York City. Its construction and operation may be understood from the cut. As in the drop-jacks previously described, the shutter is restored to the normal position by the act of inserting the plug in the jack. The plug passes through a hole in the shutter,

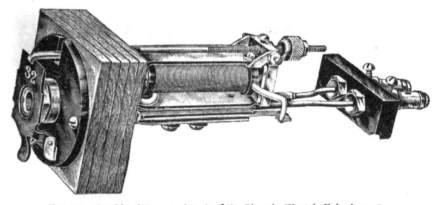

FIG. 145.—Combined Drop and Jack of the Phœnix Electric Telephone Co.

and, when pushed through, presses against the shutter, thus restoring it. The drop itself is mounted on a hard rubber disc, which, in turn, is mounted on a wooden panel, so that by removing the two screws the drop can be taken out of the board without disturbing the wiring. The two back contacts of the drop pass into large double binding posts, which are mounted on thick hard rubber strips attached to back of board. This arrangement is such that the line wires leading to these binding posts need never be disturbed, should it be necessary to remove the drop for any purpose. The drop is held in front by the two screws through the hard rubber disc; the back contacts are firmly screwed in these double binding posts.

CHAPTER THIRTEEN.

SWITCHBOARD LAMP SIGNALS AND CIRCUITS.

Lamp Signals for Switchboards.—In many exchanges line drops of all descriptions have been supplanted by small incandescent electric lamps as calling signals. These lamps are superior to drops in that they need not be "restored," the light being extinguished so soon as the proper connections are completed, and also from the fact that they occupy far less room on the panel, a consideration of especial importance in the construction and operation of multiple switchboards of large capacity. On the other hand, they present the disadvantage of giving the operator considerable physical annoyance, which, in many cases, is hardly compensated by the time and labor otherwise saved. Figs. 146 and 147 show, respectively, the usual size of such signal lamp, and the method of enclosing it in a suitably shaped cup or cell. Fig. 148 shows a row of lamps mounted on a strip for attachment to a switchboard.

Methods of Mounting Lamp Circuits.—There are several structural disadvantages involved in the use of lamp signals, and any exchange in which they are installed must be constantly inspected in order to prevent any line from being rendered inoperative by the burning out or damaging of its lamps. As a usual thing, it has been asserted, switchboard signal lamps may be flashed at least 1,000,000 times without serious damage. But, on the other hand, and particularly when the lamp is included in the line circuit, any outside electrical disturbance, such as will increase the strength of the current unduly, will cause the lamp to burn out much sooner. Usually, therefore, the circuits of signal lamps are made separate from the main ···e, and depend for operation on a series of circuit-closing ys, as will be presently explained.

Lamp Signals on the Main Line.—The most typical method of attaching the signals to the main line, between the wires of the subscriber's metallic circuit, is, briefly, to bridge the jack between the line wires and attach the lamps *in series* to one limb of the circuit. Each terminal of the line then passes through an impedance coil, or long-wound magnetic induction resistance, and is connected to either pole of a battery. This battery, while the subscriber's apparatus is in normal condition. is prevented from illuminating the lamp signal by the high resistance of the call bells, which are usually wound to a resistance of 1,000 ohms. So soon, however, as the subscriber removes his receiver from the hook, thus making the telephonic circuit, the battery current finds a ready path in series through the receiver and the secondary of the induction coil, a line of far smaller resistance, and immediately flashes the signal. Of course in such an arrangement as this the magneto-generator of the subscriber's apparatus is omitted, and the central energy exchange system adopted to the extent of deriving signal power from "central." The subscriber's sole act in signaling is to remove his receiver from its hook. The fact, however, that the central signal battery is always on closed circuit through the impedance coils at both terminals of every line, a further advantage may be derived from including both the station bells and a single-cell storage battery, of the general type already explained, in a permanent bridge between the two limbs of the circuit. Thus the local storage battery is constantly being charged to its full capacity while the telephone circuit is open; but, if at any time it become exhausted, and the line be brought into use before charging is complete, the transmitter may derive sufficient current from the central battery, through the bridge to one terminal of the induction coil primary on the one side, and through the battery and switch hook contact on the other, to operate with good effect. The connection of two subscribers' lines through a plug pair operates to extinguish the signal lamp by introducing a sufficient resistance to render the battery incapable of maintaining the

Fig. 148.—Row of Incandescent Signal Lamps enclosed in tubes and faced with glass lenses, for mounting on a switchboard panel.

Figs 146-147.—Small Incandescent Electric Lamp for a Switchboard Signal, and enclosing tube in which it is mounted on the panel.

light. At the same time the impedance coils act to balance the lines connected at the bridged jacks.

Lamp Signals Operated by Relays.—The second method of arranging circuits operating switchboard lamp signals is by a system of relays and subsidiary battery circuits. The most typical system, the invention of C. E. Scribner, of Chicago, is briefly as follows: The two limbs of the line, attached respectively to the two springs of the switchboard jack, terminate, the one in a ground connection, the other in a wire common to all lines and containing a grounded battery. As in the type of circuit already described, each subscriber's apparatus has the bell magnets on a permanent bridge, which also contains a condenser, generally of about .75 microfarad. The introduction of the condenser practically breaks the line circuit, so far as the direct current central battery is concerned, although presenting no obstacle to the alternating current from the calling magnets. It would thus permit the passage of the telephonic current, also alternating, were it not that the coils of the bell magnets were wound to a high resistance and impedance for the express purpose of preventing this short-circuiting of the talking current. As soon, however, as the receiver is removed from its hook and the telephone circuit is made, the current from this central battery finds a path of low resistance through the receiver coils and the induction secondary, and is able to energize a relay connected in series with the line limb which is joined to its positive pole. This relay attracts its armature, a pivoted bar having a grounded connection, and brings it into contact with an anvil attached to one terminal of the lamp circuit. The other terminal of this lamp circuit is attached to a wire, common to all the lamps on the board, and carrying a battery with a grounded connection. Therefore, so soon as the relay attached to the line circuit attracts its armature, it makes the circuit through the lamp from the battery just mentioned and to ground through the pivoted armature, thus causing the proper lamp to be illuminated as a calling signal to the switchboard operator. The lamp then continues lighted until the answering plug is inserted in the jack of the calling subscriber.

Jack and Plug Circuits.—The act of inserting the plug in the jack causes the lamp to cease burning, as follows: Each jack has three contacts, a spring for tip, a spring for sleeve, and a thimble connected to a third relay, which is grounded. Each plug has the usual tip and sleeve strands, and, in addition, a thimble contact, insulated from both tip and sleeve, and connected through a battery to ground. The insertion of a plug in a jack, then, makes the usual line contacts, and also makes the circuit of the third relay, just mentioned. The armature of this relay consists of two hinged metallic contacts, which normally bear on two anvils, one at the terminal of each limb of the sub-scriber's line. Consequently, as soon as the relay is energized the line connections to ground and through the first-mentioned battery are cut off, and the jack springs are left as the sole terminals of the line. Moreover, as the relay continues to hold its armature so long as the plug is in the jack, it is impossible that the lamp be relighted before the conversation is completed. Associated with the line signal lamp is a system of pilot lamps, actuated by yet other relays, to assure the operator's attention; but this feature need not be noticed here.

Both these arrangements of lamp signals belong properly under the head of the exchange battery, or central energy systems, with which they are best adapted to operate.

CHAPTER FOURTEEN.

THE MULTIPLE SWITCHBOARD.

The Requirements of Large Exchanges.—The various forms of one or two section switchboards described and illustrated in the previous chapters are intended to operate on systems having only a few subscribers—say 200 or 300—and capable of being managed by one or two operators. When there are only a few hundred subscribers wired to an exchange it is a comparatively simple matter to make the required connections, by reaching over with the cord of the calling plug and inserting the plug in the required jack, even when it is in the section of another operator. As may be readily surmised, however, such a procedure, while perfectly satisfactory in exchanges where there is very little business, are utterly impracticable when the number of subscribers has reached into the thousands, and each operator works to her full strength during certain hours. Here it is necessary to have some means whereby an operator can, with the ease and speed of handling her own 100 or 200 drops and jacks, make any desired connection, even into the thousands. Otherwise, in order to meet the requirements of rush hours, the number of operators would have to be far greater than one to each hundred subscribers.

The Arrangement of Multiple Jacks. — The device adopted to meet these requirements and, in a great measure, overcome the necessary difficulties of handling a large number of subscribers, is called the multiple switchboard. This name is strictly descriptive, and indicates the peculiarity of the system adopted, whereby each subscriber's line terminates, not in one jack merely, but in a number of jacks equal to the number of switchboard sections in the exchange. Thus each operator h

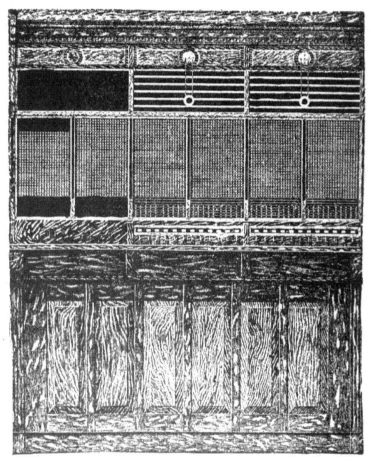

FIG. 149.—A Typical Multiple Switchboard. One section of a 3600-line Multiple Board of the Stromberg-Carlsson Co., equipped with "visual signals," line drops of the type shown in Fig. 162, and operated by full central energy at the exchange. The left-hand position is left uncompleted, but, as may be seen, every line is "multipled" in the three positions shown, so that the middle operator may reach any subscriber on the line, either on her own panel or on the one to her left or her right. Below each multiple panel may be seen the usual 100 "answering jacks" having drops in that position.

on the panel before her, not only the jacks corresponding to her own subscribers' drops, but jacks connecting with the lines of every other subscriber in the exchange. Where there are as many as 5,000 or 6,000 subscribers, as in the largest exchanges, it is obviously impossible that so many jacks should be arranged in any one panel; consequently the number is arranged to occupy, say three panels, so that every number of the 5,000 or 6,000 is repeated at every third panel. Thus each operator, either by inserting the plug in the jack of required number on her own panel, or by reaching with the cord of her calling plug to the panel on her left or to the one on her right, can make any desired connection out of the many thousands registered in that exchange. This arrangement is shown in Fig. 149.

Varieties of Multiple Switchboard. — The method of arranging the wiring of a multiple switchboard so as to accomplish this *multiplication* of the jack connections of each subscriber is, briefly, to run each line the whole length of the switchboard, instead of having it end at the panel where its drop is fixed, and to "tap" it for jack connections at every panel, or every three panels, as just explained. There are two ways of attaching the jacks on a line—either in "series" or in "parallel." The latter method is known, in telephone parlance, as the "branch terminal" wiring, also the "three-wire system," and is, briefly, the method of bridging the circuit between the line and return wires of each subscriber over as many jacks as the system requires. It is the system now adopted in most up-to-date exchanges.

The Series Multiple Switchboard.—In Fig. 150 is shown the plan of wiring the jack and drop connections of a series-wired, metallic circuit multiple switchboard of 300 drops. Three out of the three hundred subscribers' lines are shown, number 45 having the drop in the first section, 110 having the drop in the second section, and 216 having the drop in the third section. Thus, Section I has subscribers' drops 1 to 100; Section II, drops 101 to 200; Section III, drops 201 to 300. Each drop

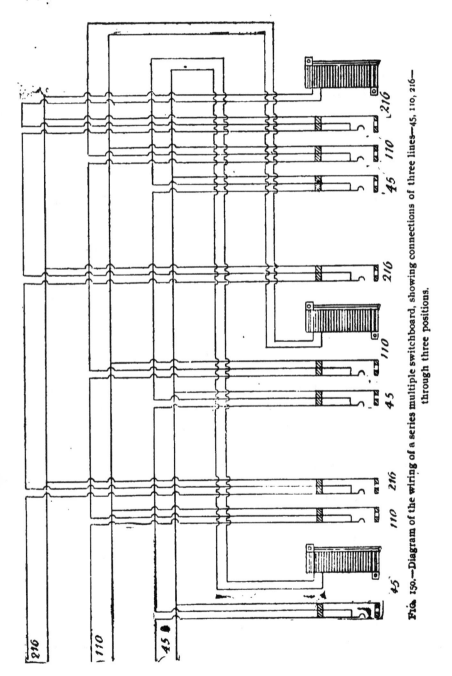

FIG. 150.—Diagram of the wiring of a series multiple switchboard, showing connections of three lines—45, 110, 216— through three positions.

has the answering jack to correspond, and the operation of receiving a call and connecting a subscriber is the same as that in the standard type of switchboard already described. In addition to the 100 drops and answering jacks belonging to each section, however, there are also 300 multiple calling jacks, representing, as shown by the above figure, every subscriber's line which is wired to the board. So in case number 45, in the first section, wishes to converse with number 216, in the third section, there is no need to communicate with the operator at three, but simply to insert the calling plug in the multiple jack belonging to line number 216, which is wired to Section I, as shown.

Testing Arrangements of a Multiple Switchboard.—In order to attain the end for which a multiple switchboard is designed—the placing of calling jacks for every subscriber, even to 5,000 or 6,000, within reach of every operator in the long row of switch panels—it is necessary that there should be some ready method for determining, at any section of the 50 or 60 in the row, whether any one line is engaged or not. To accomplish this end a still further complication of machinery is necessary. An extra battery and a grounded circuit are bridged between the sleeve strand of the calling plug and the operator's telephone set, for the purpose of testing the lines, in the manner to be presently explained.

Test Circuits of a Series Switchboard.—Fig. 154 shows the sleeve, test and listening connections of a series metallic multiple switchboard. In order to avoid the bewildering complications, unavoidable in the use of the ordinary diagrams, the circuit terminals are represented as connected with a type of listening and ringing key, which has already been fully explained, both as to its construction and operation. In this figure the important points are the sleeve strand and the listening circuit. Consequently the other connections of the switch key are merely indicated. Here *A* is the wire connected to the answering plug, and *C*, that connected to calling plug, of a given pair, the tip

FIG. 151.—Strip of Answering Jacks.

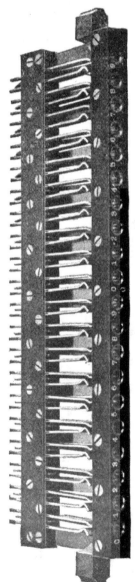

FIG. 152.—Strip of Multiple Jacks. These two strips are the same length, intended to show the differing arrangements as applied to the jacks having drops in a given section or those that are "multipled" through the full length of the board.

FIG. 153.—Plug for a metallic circuit multiple switchboard, showing the three insulated contacts, tip, sleeve and collar, to fit the jacks, as described. These three figures illustrate parts of the multiple switchboard construction of the Western Telephone Construction Co.

and sleeve strands being connected to the key, as already
explained. To the sleeve terminal of *C* is connected a battery
of cells, *B*, connected to ground at *X*. As they are permanently
bridged across the sleeve strand of the plug circuit, this battery
and ground connection would undoubtedly unbalance the line,
with the probable result of short-circuiting the talking current
with some other connected line, were it not that the impedance

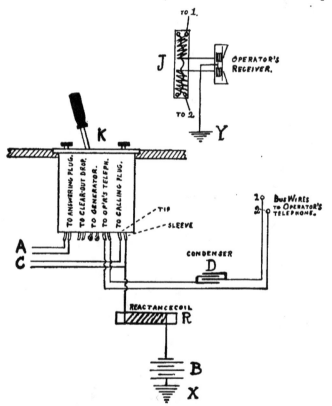

FIG. 154.—Diagram of the test circuit of a series multiple switchboard, simplified
by using a familiar form of switch key.

or reactance coil, *R*, is placed as shown. This coil offers a much
higher resistance than the telephone line, and, as a result, effectually prevents all shunting or interference. Consequently
when the cord circuit is connected, and the battery, *B*, left in

connection with the line, as it is easy to see must be the case, the working of the telephonic current is in no way interfered with. The battery, *B*, however, is intended to act only when some other operator's telephone set is bridged into line with the tip of her calling plug, as will be afterward explained. Interposed on the line leading to the operator's receiver will be seen the condenser, *D*, which is intended to prevent disturbances in the line from producing conditions liable to give a false " busy " test. Furthermore, the secondary winding of the induction coil is split into two parts, one end of each being connected to line with the plug strands, tip and sleeve, the others with the pole coils of the receiver magnet. The ground connection shown at the receiver leads from the center point of the magnet coils, so that such electrical impulses can be run from the battery, *B*, through the condenser, *D*, pass through one side of the secondary winding of the induction coil, and through one side of the magnet coils to ground.

Method of Testing.—By reference to Fig. 150 it will be seen that the series wiring to the jacks is made only from the line wire, the return wire being strung parallel to it along the entire length of the board, and connected "in multiple" with the thimbles provided to give the sleeve contacts of the plugs. The effect, therefore, of inserting a calling plug in a multiple jack is to connect the thimble, or sleeve contact, of every other multiple jack on that subscriber's line with the battery, *B*, through this return, or test, wire. As soon as a circuit is formed to ground from the battery, *B*, by making contact with the thimble of any multiple jack in a plugged line, an electrical effect must follow. The method of making the test is as follows: Suppose that subscriber 10, in Section I, wishes to communicate with subscriber 326, in Section IV, he turns the crank of his apparatus generator, thus calling the attention of the operator at Section I, who inserts an answering plug in his answering jack. Having ascertained his wishes she must make a test to discover whether line 326 is plugged at any other section of the switchboard, and to do so holds her telephone set on circuit, and touches with the tip of the calling plug, belonging to the same pair, the thimble of the multiple jack of subscriber 326 on the panel in front of her. If subscriber's line 326 is engaged, she will receive the "busy test" as follows: The tip of the plug, touching the thimble of multiple jack 326, will close a circuit with the grounded battery at 326, Section IV, through the test wire, as shown to the thimble of the multiple jack, numbered 326, in Section I; thence through the tip strand of the plug, through the condenser, *D*, through one-half of the secondary winding of the induction coil of operator, Number I, through one pole coil of her magnet receiver, through the ground wire connected to the middle point of coil wire, to earth and to the ground connection of the test battery in action at Section IV. The condenser, *D*, being placed in the line, while preventing the transmission of a continuous current, will discharge into circuit with the result of giving a sharp clicking sound in the operator's receiver, thus informing her that the line is busy at some other section. If the line is not busy, no result whatever will follow the act of touching the thimble of the multiple jack with the plug tip, and connections may be made at once.

The Use of the Condenser in a Test Circuit.—While the electrical condenser, like the induction coil, forms a complete break in the circuit, so far as a continuous current is concerned, it is useful in a variety of ways in both telephony and telegraphy. Even though the plates composing a condenser of ordinary pattern be insulated from one another, the rapid variations and reversals of the telephonic current produce an inductive action between them, and allow the transmission of speech with ease and perfectness. It is also the device best suited to the test needs of a multiple switchboard of the series type.

Defects of a Series Multiple Switchboard.—The series multiple switchboard is open to criticism in a number of particu-

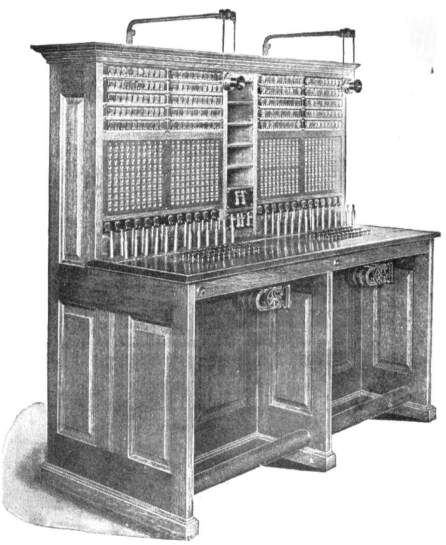

FIG. 155.—Couch & Seeley's 200-drop multiple switchboard. This cut well illustrates the multiple system. showing 400 jacks to 200 drops, or each line carrying two jacks, thus avoiding cross-connections.

lars, most prominent among them being the constant liability to disarrangement of a line by a particle of dust or some foreign substance lodging between the springs of a multiple jack and insulating them, and the difficulty in using self-restoring drops on the lines, as would be a most desirable addition in the complicated machinery of a multiple switchboard.

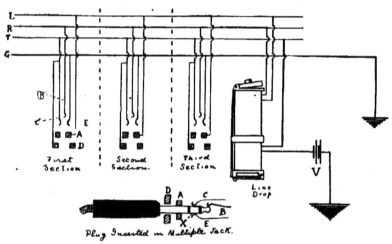

FIG. 156.—Diagram of one circuit through three positions in a branch terminal multiple switchboard.

To remedy these and other grave difficulties, the other type of multiple board, the parallel, or multiple-wired, branch terminal, or three-wire board, was devised. While much more complicated in some respects, and requiring more appliances for its successful operation, this type of switchboard is coming into more extended use.

The Branch Terminal Multiple Switchboard.—Fig. 156 shows the wiring plan of one type of branch terminal multiple switchboard. For convenience, only one line is shown through its multiple connections in three sections. Here are the two line wires, as in the series board, marked *L* and *R;* also a third, the test wire, *T*, running parallel to the other two through the whole length of the board, and the common ground wire, G, which, unlike the test wire with which it is connected when any test circuit is closed, is common to all the jacks in the board, belonging to all subscribers whatsoever. The jacks of this type of board have five contact points—three springs and two thimbles, as shown. The thimble, A, is connected to the wire, L; the short spring, B, to the wire, R; the spring, C, and the test thimble, D, to the test wire, T, and the spring, E, to the common ground return wire, G. Self-restoring drops, of the type already described, are used with this board. Thus, as may be seen, the insertion of a plug in any jack closes two circuits—the telephonic circuit, on metallic line, and the drop-restoring circuit, on grounded line; besides leaving the grounded battery connected to the test thimbles, D, ready to give the necessary "busy" test at any section of the board, in the manner already described in connection with the series multiple board. In order to make those connections, the plug of a three-wire switchboard must be of slightly different construction from the ordinary type. This is shown in Fig. 156, where is represented a plug inserted in a jack of a three-spring variety. As may be seen, the tip of the plug registers with spring B; the sleeve with thimble A; the springs, C and E, are forced apart by the entrance of the plug, and are held in electrical connection by the metal collar, X, which is fitted over the shank of the plug, as shown, but separated from it by an insulating ring. The test thimble, D, is made sufficiently wide to permit the insertion of the plug without making contact.

The operation of the speaking circuits of this type of board is the same as in all the types of boards previously described. The circuit of the self-restoring drop is made, as shown by the connection of the two springs, C and E, instead of by the connection of two thimbles, as in the standard type of board, when strung for these drops, as already described. The drop-restoring and test battery, V, is, like the ground connection, G, of the test wire, common to all subscribers' lines.

The Branch Terminal Testing Circuit.—The test circuit of a branch terminal board is constructed somewhat differently from that of a series board, but the manner of making the test is the same, the "busy" signal is the same, and, as in the other type of board, it is switched into operative connection when the table telephone set is bridged into line. Fig. 158 shows the general features. Here K is the switch key, already described; A and C, the connections to the answering and calling plugs,

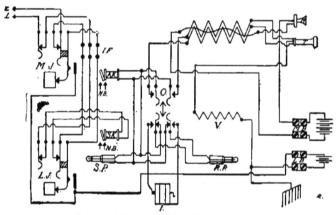

Fig. 157.—**Diagram of the multiple switchboard shown in the frontispiece.** L is the Line; $I F$, Intermediate Field; O, Table-Key; $M J$, Multiple Jack; $S P$, Speaking Plug; $N B$, Night Bell; $L J$, Home Jack; $R P$, Ringing Plug; V, Busy-test Coil; $S K$, Subscriber's Drop; I, Generator; $L B$, Test Battery; $A K$, Clearing Out Drop; M, Microphone; $M B$, Microphone Battery; T, Telephone. It will be seen that the jacks used on this board are of the type shown in Fig. 160, and that the insertion of a plug throws the battery current into all the multiple thimbles, the sleeve spring, when actuated closing connection with the test battery at $L B$.

respectively; D, the terminal of the clearing-out drop, here of the self-restoring variety; G, the terminals of the ringer generator; and T, the terminals of the operator's talking set. As will be seen, the wires leading from D are connected direct to the operating coil of the clearing-out drop, which is permanently bridged between the plug cords when talking connections are made.

As in the test circuit of a series board, the secondary winding of the operator's induction coil, J, is divided into two halves, two ends of which connect to the terminals at T, and the other two to the poles of the receiver. Here, also, the

ground connection is made at the middle point of the receiver
coil wires; with the marked difference, however, that the
battery, *B*, is interposed, thus making the test battery circuit
in this board the reverse of that in the other type. Of the two
terminals of *T*, one is connected by a branch wire, as is shown,
direct to ground through the common ground wire, *Y*, thus
making the two terminals of the test circuit, from ground at *X*,
through battery *B*, to the middle point of the receiver coil,
through one-half of the coil, through one-half of the secondary
winding, at *J*, to ground at *Y*.

Making a Test.—In order to make the "busy" test the
operator works the key, as formerly explained, so as to bridge
her talking set across the strands of the calling plug. She
then touches with the tip of this plug the thimble of a multiple
jack belonging to a subscriber who is called at her section.
If his line is disengaged no electrical effect will follow, and
connections may be made at once. This is true because, when
the line is idle, the test ring and the tip of the plug, both
connected to ground through the battery, *B*, are at the same
potential, and no current will flow. By the connection of the
springs, *C* and *E* of the jack, however, by the insulated collar
of the plug, a difference of potential between the terminals, plug
tip and test thimble and ground is created, and on the contact
of tip and thimble an active circuit is made, with a resulting
click, which may be heard in the operator's receiver.

When the operator's talking set is bridged into the cord
strands of the plug another result is produced. Connected to
terminal 2 of *T* on the switch key, will be seen a line leading to
one side of the restoring coil, *R*, of the clearing-out drop.
This enables the closing of the drop, should it be open, by the
simple act of switching in the talking set, an arrangement
having its advantages when we consider that it is usual to place
self-restoring drops behind glass, as protection from dust and
other interferents. With the bridging-in of the talking set, therefore, a circuit is formed from ground at *X*, through battery *B*,

through the restoring coil, *R*, through one-half of the secondary winding of the induction coil, *J*, through both pole coils of the receiver, through the other half of the induction secondary, to terminal 1, and thence through its connected line to ground at *Y*.

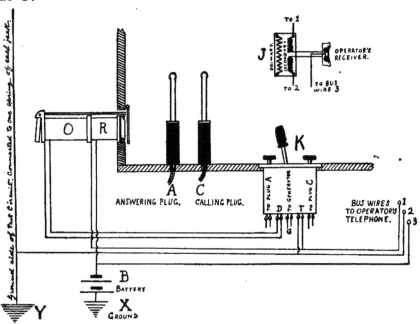

FIG. 158.—Diagram of the test circuit of a branch terminal multiple switchboard, simplified by the use of a familiar form of switch key.

Visual "Busy" Signals.—A number of multiple switchboards, more particularly such as operate by a common exchange battery, are equipped with a "busy" signal system of a different description from any of those mentioned above. It consists, in brief, in arranging a small electro-magnet and drop shutter in connection with each multiple jack, so that when a given line is plugged, a local grounded battery is thrown into circuit, and all its multiple jacks are closed by drop shutters, which enables the operators to tell at a glance whether or not the line is busy. Such an arrangement, of course, is advantageous in obviating the necessity of making the "busy" test, as described above, and also in simplifying the wiring of the multiple jacks. Further it does away with the complicated test circuit arrangements, which must involve added perplexity in the operator's cord attachments. It has not, however, come into general use.

Divided Multiple Switchboards.—The vast complication of the ordinary type of the multiple switchboard has been the occasion of numerous inventions proposing to simplify the necessary plant of a large exchange. The most representative of these devices are, respectively, the divided multiple switchboard, and the various locally inter-connected, transfer or express systems, whereby *trunking plugs* are used to connect the several positions of a switchboard, enabling the making of connections without the use of multiple jacks. The divided switchboard is the invention of Milo G. Kellogg, of Chicago, and, as manufactured by the company of which he is the head, may be briefly described as follows: In an exchange of, say, 6,000 subscribers, which means 60 positions of 100 drops each, and at least 20 multiple jacks to every line, we have 120,000 jacks at least, exclusive of answering and trunking jacks. Mr. Kellogg's system accomplishes the very desirable result of quartering this figure, giving a total of say, 30,000 multiple jacks to a system of 6,000 subscribers. This he does by dividing the 60 positions mentioned into four divisions, and constructing the calling apparatus with polarized drops and ordinary pole changing switches.

Polarized Drops.—Let us suppose that each of the four divisions of the Kellogg switchboard represents an even 1,500 subscribers. Then subscriber No. 120, whose drop is at position 2 of division 1, has a drop not only in that position, but also at a corresponding position in divisions 2, 3 and 4. By this device he may call any subscriber he desires in either of these three divisions. The device is simple. His magneto-generator, or other source of calling current, is furnished with a pole-changing switch, which may be so manipulated as to enable him to send a positive or a negative current over the line wire with a ground or third-wire return, or a positive or a negative current over the test wire with a ground or third-wire return. The first of the four currents mentioned will energize, say, his drop magnet in division 1; the second, his drop in division 2; the third, in division 3; the fourth, in division 4, each of these drops being so polarized as to respond to that particular current and no other. Thus he may transmit a call to any one of the four, as he desires, according to the position of the subscriber with whom he seeks communication.

Answering and Multiple Jacks.—Subscriber 120 has an answering jack in position 2 of division 1, as we have seen, and also in the corresponding position in divisions 2, 3 and 4. In division 1 his line is also tapped with multiple jacks at every section in the ordinary fashion. But he has no multiple jacks in the other divisions, the sole object of connecting his line to them being for purposes of calling subscribers belonging there. Any other subscriber, therefore, who wishes to communicate with 120 will so arrange his calling circuit, by his pole-changing switch, as to release his drop in division 1, where a multiple jack of 120 may be plugged as desired.

Advantages of the System.—In addition to those arrangements there are also relays which operate to break the circuits of all the drops of a subscriber the moment a plug is inserted in any jack on his line, thus obviating the confusion arising from false calls. The testing circuit is arranged in the same fashion as in the ordinary multiple switchboard. By the use of the divided board system, not only may the number of multiple jacks be reduced, but four times as many subscribers may be wired to an exchange. The limit on an ordinary board is 6,000 subscribers, with 120,000 jacks. With a Kellogg divided board exchange there may be 24,000 subscribers, and yet but 120,000 jacks, on the principle just explained. Such an arrangement, if successfully carried out, must mean an increased activity in the telephone business and a corresponding reduction in tariff rates.

FIG. 159.—An Independent Telephone Exchange at Newark, N. J., equipped with the multiple transfer system of the Western Telephone Construction Co.

CHAPTER FIFTEEN.

LOCALLY INTERCONNECTED OR MULTIPLE TRANSFER SWITCHBOARDS.

Objections to Multiple Switchboards.—While the multiple switchboard is a very complete and in many ways a satisfactory device for accomplishing the ends for which it is constructed, its great complexity, and the correspondingly increasing cost of installing and enlarging exchanges equipped on the multiple system have set numerous inventors to work on the line of devising satisfactory substitutes. Their efforts have resulted in the several systems of transferring, by which the line of a calling subscriber may be directly switched to the board on which is the jack of the called subscriber. While in all these systems there is required very complex appliances, the absence of the multiple jack feature more than compensates for it.

The Sabin-Hampton Express System.—Perhaps the most typical, and certainly the most complicated, of the transfer systems is the Sabin and Hampton "express" system. This operates, briefly, as follows: In the first place, the apparatus at the subscriber's station is constructed without the ordinary magneto generator, the operation of the drop, or visual signal, on the switchboard panel being accomplished by the current from a storage battery at the exchange, by closing a circuit from the subscriber's station to the exchange, on the removal of the receiver from the switch hook, as in the operation of the common battery systems now increasing in popularity.

Cut-Out Jacks.—Each subscriber's jack contains two spring contacts of the ordinary type, which form the terminals for both sides of the line—the line wire and the "test" wire—and each of them rests, when not actuated by a plug, on an anvil, from

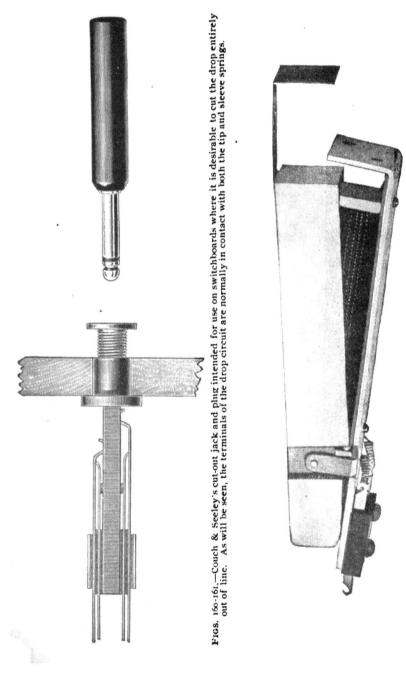

FIGS. 160-161.—Couch & Seeley's cut-out jack and plug intended for use on switchboards where it is desirable to cut the drop entirely out of line. As will be seen, the terminals of the drop circuit are normally in contact with both the tip and sleeve springs.

FIG. 162.—Western Rising Visual Signal for use on central energy switchboards. The magnet coil is energized by a direct current battery after the circuit has been made with the subscriber, and the target remains raised until the plug is inserted in the cut-out jack.

which the plug draws it away in precisely the same fashion as in the types of jacks already described. The jack of this system is, in fact, a double jack, having two line springs and two drop contacts. A jack of this description is shown in Fig. 160.

Rising Visual Signals.—When a subscriber takes his receiver from the hook the circuit is closed at his apparatus, through the two drop anvils of his jack, through the local battery and the coil of the drop, causing the armature to be attracted and the target to be raised so long as the plug is not inserted in the jack. Immediately this is done, the circuit is broken at the jack anvil contacts, and the target falls into its normal position. Such a line signal is shown in Fig. 162.

Call Boards and Order Boards.—There are two kinds of switchboards required by this system, which for convenience, we may designate as the "call" boards and the "order" boards, respectively, although these terms are here adopted merely for simplicity's sake. The "call" boards are of the ordinary type, containing the signals and jacks of 100 subscribers to a panel, as in all others, but without the operator's listening and ringing keys already described. The "order" boards are equipped with signals and jacks corresponding to every section of the call board, so that connections may be made between the two.

Operating the Call Boards.—The plugs of the call boards differ in two particulars from the ordinary type. First, they are arranged in single rows instead of double, as in other boards; second, they form the terminals of a "trunking" circuit between the call boards and the order boards. That is to say their tip and sleeve strands connect direct to the signals—each plug has one—on the order board. Thus when a subscriber calls, and his signal rises at the call board, the operator there merely inserts a "trunking" plug in his jack, causing his signal to be restored, and pays no further attention to it.

Operating the Order Board.—The insertion of the plug in the jack of the calling subscriber continues his line from the call board to the order board, where the proper signal is displayed. The operator at this board then moves the proper switch key, so as to bridge in her telephone set and inquire the wants of the subscriber thus transferred from the call board. Having ascertained the number of the subscriber with whom he wishes to converse, she presses the "order key" of a trunk line leading to the call board section containing the called subscriber's jack, and by this act having placed herself in direct connection with the telephone of the operator there, directs her to make the proper connections. The operator at the second call board, thus signaled, answers, informing her which one of the several trunk plugs to use; and this being complied with makes the desired connection.

Lamp Signals and Relay Circuits.—The signal and clearing devices used on this type of exchange boards is at once ingenious and complicated. Instead of drops of any description incandescent electric lamps, white and red, are employed for this purpose on the order boards. The trunking plug being inserted in the jack of the calling subscriber at the first, or incoming call board, completes the circuit closed at the subscriber's station by the removal of his receiver from its hook, from its tip strand, through a balance coil arranged to prevent interference between the signaling and talking currents, through a local battery, which energizes the coil of a relay, constructed on the same plan as the ordinary switchboard drop, thus causing it to attract its armature, and from this coil to the sleeve strand of the trunking line. When the relay attracts its armature it closes another circuit between a battery and the signal lamps on the table of the order board. As soon as the circuit is thus made the white lamp is lighted, and continues burning until the operator of the order board switches in her telephone set; or until she lifts the plug from its socket, to insert it in the jack of the section containing the called subscriber; or until the calling subscriber replaces his receiver on its hook, thus breaking the circuit.

Operation of the Signals.—The withdrawal of the plug from its socket at the order board, in order to insert it in the required jack, and thus make the trunking connections with the outgoing call board, releases a leaf spring, which breaks the circuit of the white light and makes the proper contacts for the red light, which is lighted as soon as the calling subscriber restores his receiver to its hook. Thus the white light acts as a calling signal and the red light as a clearing-out signal. Remembering that the red lamp depends for its proper connections on the removal of the plug from its socket, we may understand how that the removal of the plug from the jack at the order board, before conversation is ended, will again make the circuit of the white light, thus causing it to burn again, as a signal to the operator that she has made a mistake. Similarly, should the operator at the first, or incoming, call board remove the plug from a jack before the conversation is ended, the calling signal, already described, would again be displayed, thus indicating that the circuit is still made from the subscriber's apparatus. The order board relay, already described, being wired in multiple to a relay at the incoming call board, operates a similar sort of clearing-out signal there.

The Clear-out Signal System.—The clearing-out mechanism of the outgoing line, as operated on all call boards whatsoever, when plugged for an outgoing line, is somewhat complicated, although highly efficient. Briefly it consists of three different relays, the combined calling and clearing drop and two others, which, by their individual energizing at the successive acts of the operator, can make or break the circuit of a signal lamp placed at a convenient point for indicating the state of a particular line. The connections of the calling and clearing relay signal are made as already described. The two other relays form a pair, one placed above the other, the poles in opposite directions, in such fashion that what corresponds to the shutter of ordinary drop coils with the upper one is the armature of the lower one. This shutter-armature normally leans away from both coils, but having been attracted by the lower of the two and then released, it is caught and held by a hook at the end of the armature lever of the upper coil, just as the shutter of the ordinary drop is held in place. This contact, however, makes a circuit which lights the signal lamp, which circuit is broken so soon as the coil of the upper relay is energized, thus raising the armature lever and allowing the shutter to fall down to its normal position.

The Signal System on an Outgoing Line.—The operation of the signal system of a call board with an outgoing line is as follows: The operator at the call board having made the connection as directed by the operator at the order board, having inserted the trunking-in plug into the designated jack, pushes the button of her ringing key, thus sending a calling current to the subscriber desired. The ringing key is so arranged that the act of pushing the button does two things—sends a current and makes the circuit between a local battery and the lower of the two relays just mentioned, causing it to attract the armature or shutter. As soon as she releases the button of the key the circuit is broken, and the armature shutter falls away from the pole of the lower relay, being caught by the lever hook of the upper relay, and thus closing the circuit from a local battery, through the shutter armature to the lever hook, through the lever to the armature pivot of the upper relay, through the incandescent signal lamp just mentioned, and back again to the other side of the battery. The lamp thus lighted by the depression of the ringing key remains burning until the called-for subscriber takes his receiver from his switch hook. This act, as with the calling subscriber, in the first instance, causes the calling coil to operate and raise its shutter, as previously noted. This raising of the shutter also makes a circuit through the coil of the upper relay, causing it to attract its armature and raise its armature lever, releasing the shutter from the hook, and by allowing the shutter to fall away again.

breaking the circuit of the signal lamp. The shutter of the calling and clearing signal, thus raised, remains so until the conversation is ended, the falling of the shutter being the clearing signal. Since the initial act of operating this system of signals is pressing the ringing key, very little reflection will show that it could not be used except on the outgoing line.

Advantages of the Express System.—Although the mechanism of the express system may seem uselessly complicated, and there are several disadvantages in the delay entailed in having three operators to handle every one call, yet the great expense and constant attention to involved circuits and delicate contrivances entailed in the installation and maintenance of a multiple switchboard are obviated. The express transfer system is much more readily operated, and the constant danger that a careless or inefficient operator will plug the multiple jack of a busy line is completely done away. This system is also contrived to save manual labor and the attention of the operator to the last degree. Instead of compelling her to answer each individual subscriber to inform him when connections cannot be made, power-driven phonographs are attached to the boards which constantly repeat into a transmitter, mounted with induction coil and battery, the required sentences. These may be connected to the line by inserting a plug in a jack connecting to the one or the other phonograph, and thus allowing the sentences: "Subscriber called for does not reply," or, "Busy; please call again," to be transmitted to the calling subscriber. The latter phonograph is attached to the order board, and is switched into line by the operator there as soon as she has learned from the operator at the second call board that the required line is in use; the former is attached to the call board, and is switched into the line of a would-be outgoing subscriber so soon as, by the lamp signals already described, the operator has learned that the called subscriber does not respond. These phonographs are useful, as enabling an operator to give the required information by a simple manual act, without interrupting her other duties to speak along the line.

The Cook-Beach Transfer System.—Another transfer system based on some of the same general principles, although employing only one type of board for both calling and order-

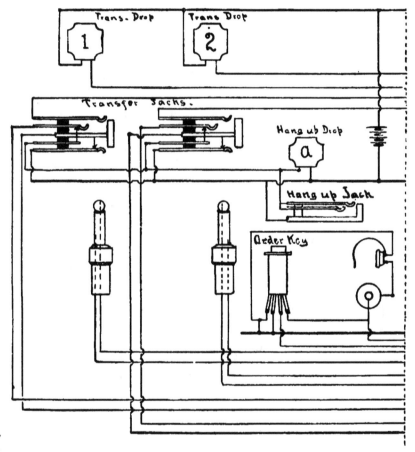

FIGS. 163-164.—Diagrams illustrating the circuits of the Cook-Beach Transfer System through two operators' sections. The transfer drops indicate on one section of the switchboard that the corresponding transfer jack has been plugged on another to which it is connected. The operator, noting the fall of a transfer drop inserts the plug connected through its strands with the signaling section into the jack of the called subscriber, whose number she learns by depressing her "order key," thus throwing her telephone set into

ing, is the Cook-Beach system, which is used in connection with the apparatus of the Sterling Electric Co. The plan is to have all subscribers' drops arranged on section panels of one hundred each. In addition to the regular switchboard apparatus—drops, jacks and plugs—there are also transfer drops, jacks and plugs, directly connected to each other several section of the board.

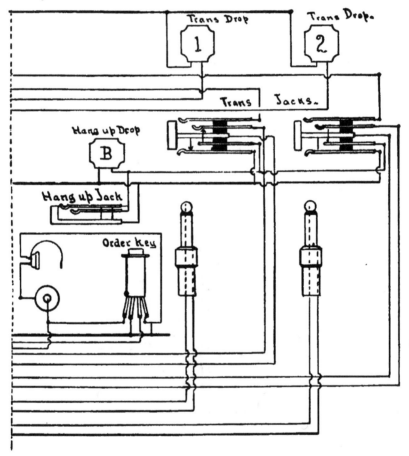

circuit with that of the operator at the signaling board. The "hang-up drops" fall to designate a disconnection made too soon, and the plug is then to be inserted in the corresponding "hang-up jack," so as to restore it. The circuit connections may be understood from the figures by remembering that all opposite lines are supposed to be continuous between the two. As may be seen, the transfer plugs at each section form the terminals there of a transfer drop and jack at the other. This is the "trunking plug."

Thus there can be both visual signaling between the several operators, and also trunking connections between the several sections of the switchboard.

Making a Connection.—When any subscriber, say Number 50, "rings up," his drop falls on his section of the switchboard, Section I, thus signalling the operator in the usual
manner.   She immediately inserts in the jack bearing his number the answering plug of a given pair, which, as we have
previously seen, is shorter, than the calling plug, the latter
being made longer, so as to act on the jack springs and cut
out the drop coil.   Having ascertained that Number 50 wishes
to converse with Number 600, the operator inserts the long
calling plug in the transfer jack corresponding to Section V of

Fig. 165.—Operator's order key used in the Cook-Beach transfer system to connect the
talking set of one section into circuit with that of another.

Fig. 166.—Cut-out plug used in Cook-Beach transfer system to break a false
connection with a busy line.

the board on which are hung the drop and jack of 600.   This
causes the "transfer drop" of 1, on V, to fall.   At the same
time she depresses her order key, which throws her talking set
into circuit with that of the operator at the fifth section, and
informs her that 50 wishes to be connected to 600.   The operator at V then takes the trunking plug on her board corresponding to the line that has just signalled, and inserts it in the jack
of the called subscriber, Number 600, thus completing connection
by the use of three plugs.

   Operation of the Signal System. — The   clearing-out
signal system on this type of transfer board is simple.   As the

drop of the calling subscriber, left bridged across the talking circuit, is usually employed for this purpose on the boards of the Sterling Company, its fall indicates the close of the conversation to the operator at Section I. As soon as she notes this signal, she removes the plug from the jack, which act causes the transfer drop on Section V to fall, signaling the operator.

If at any time before the close of the conversation, either operator should remove the plug from the jack, the act causes the shutter of the "hang-up drop" on her section to fall, thus apprising her that an error has been made. She can then insert her plug in the "hang-up jack," and learn from the subscriber what his number is, and then remake the connections. On the other hand, if a calling operator should make the mistake of making trunking connections with a line that is already busy, the operator at the called board may, by depressing a cut-out button on her transfer key, break the connection.

The Western Express Transfer System. — The transfer system of the Western Telephone Construction Co., of Chicago, is simpler in many respects than either of those previously described. On the table of each 100-drop section are arranged twelve pairs of plugs, one of each for calling and the other for answering. Ten of these pairs have bridged between the two plugs a trunking line, which leads out from that section, say, from the first, to the fourth, seventh, tenth sections—to corresponding plugs at every third section. Each transfer plug represents a distinct wire, so that there are ten distinct trunking lines leading to and from every section of the switchboard. Moreover, the transfer line of each pair of trunking plugs is attached midway between the calling and the clearing-out drop in every case, so that the clearing signal may be given at the section containing the drop of the calling subscriber, while the drop at the transfer section is cut out. A board equipped with the transfer apparatus of this system is pictured in Fig. 139.

Operation of the System.—On receiving a call at any section, the operator merely inserts the answering plug of a given pair in the corresponding jack, after the usual manner, and, having switched in her telephone set, inquires his wishes. If he asks to communicate with another subscriber at the fourth, seventh or tenth board, she depresses a push button on the table in front of her, which makes connection with the talking set of the operator at that particular board, and informs her that a subscriber on the first, second, third, fourth or other trunk wishes to communicate with one numbered on her board. The operator, thus addressed, takes the calling plug of the designated pair, and, inserting it in the jack of the called subscriber, makes the connections in the usual manner. It will readily be seen that but two plugs are used in any connection, even to the end of the switchboard. If the called-for subscriber stands in any other section than one that is third, sixth, ninth or twelfth, etc., from the section of the calling subscriber, the operator at his section calls up the operator at the section nearest to that containing the desired number, which will be either to her right or to her left, and she inserts a plug in his jack.

The Clear-out Signals.—The clearing signals of this system consist in a series of incandescent electric lights controlled by relays whose circuits are made by spring contacts arranged at the base of the answering and of the calling plug of each pair. When, as we have seen the shutter of the clearing-out drop falls at the section of the calling subscriber, she removes the answering plug from the jack, and replaces it in its socket on the table in front of her. As soon as it is thus replaced it depresses by its weight a contact spring, thus making a circuit through a relay and a local battery to a miniature lamp at both that board and the one of the called subscriber. The lamp thus lighted at the latter section informs the operator there that the conversation is completed, and she removes the calling plug from the jack of the called subscriber and replaces it in its socket. By this latter act she allows it to depress a contact spring, which breaks the circuit of the lamps and informs the operator at the first section that the disconnection is made.

CHAPTER SIXTEEN.

EXCHANGE BATTERY SYSTEMS.

Advantages of a Common Battery.—In conducting and maintaining an extensive exchange system it is, of course, extremely desirable that the details should be, as far as possible, simplified, and that all points of avoidable expense and care should be eliminated. Such ends are very largely attained by the use of the various common battery exchange systems now being so widely adopted. The simplest application of this principle consists in using a common calling battery for all subscribers' stations, the signal being given at the switchboard by closing a circuit on the removal of the subscriber's receiver, thus dispensing with the ordinary magneto-generator. Such an arrangement has been described in connection with simple lamp signal circuits, and also the Sabin-Hampton "express" system, which, as far as the subscriber's calling circuit is concerned, are fairly typical common battery systems. As in both of these systems the insertion of a plug in the subscriber's jack cuts out the signal and makes the circuit for the talking battery. A further elaboration consists in attaching a constant current battery to the cord circuit of each pair of plugs, so that, when the signal battery is cut out by plugging the jack, energy is supplied for operating the transmitters. These general principles are largely typical of all the various systems of common battery exchanges, although the devices employed for their successful application in practice vary widely.

Signaling Circuits and Batteries.—Fig. 167 represents a common type of central energy subscriber's station apparatus, the practical operation of which must be obvious to any one. The transmitter and induction coil are attached to the head of the backboard, and the ringer and switching apparatus within

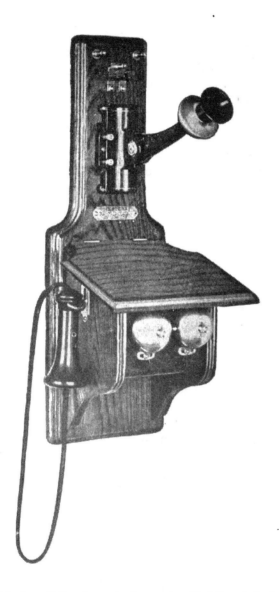

FIG. 167.—Telephone Station Apparatus to operate with full central energy. As will be noted, the magneto generator and battery box are both omitted; the calling signal being transmitted on the rising of the switch hook which also makes the circuit of the telephone set.

the box below. No battery whatever is included. To signal central *from* such an apparatus the subscriber merely unhooks the receiver, with the result of closing a circuit with the switch-board signal, as has been already described. This result is accomplished generally by making a path of small resistance in place of one of large resistance, which is maintained so long as the apparatus is not in use. Such a resistance is obtained by a high winding of the ringer magnets—they are sometimes wound as high as 5,000 ohms—which prevents an efficient current from reaching the switchboard signal. As soon as the receiver is unhooked there is provided a path of comparatively no resistance along which the current can readily flow from the exchange and back again, so as to energize the signal coil or flash the call lamp. The result is also accomplished by closing a normally open circuit between the two limbs of the line.

Calling a Subscriber.—There are two general methods of connecting a subscriber's ringer to the exchange switchboard; either by connecting it *in series* on a bridge between the two wires, the high resistance of its magnets barring the current as soon as the telephonic receiver and transmitter are switched into circuit; or else by placing it with one side normally connected to the test wire of the circuit and the other grounded, in order to enable it to take current from a grounded battery at central, the talking apparatus being normally cut out of circuit by the depression of the switch hook. In both cases the ringing circuit is made by the insertion of the calling plug in the line jack, the apparatus talking set being meantime out of circuit, as above explained. According to these arrangements of circuits the apparatus is normally in condition to receive a call, and can transmit a call by the simple act of raising the switch hook. These variations from the ordinary scheme of apparatus arrangement are necessitated by the new conditions introduced by the use of a central battery, in accordance with which problems arise on the point of how to use a central energy current to advantage without having it too weak to energize the switchboard signal or too powerful for use on the **talking** circuit.

Sources of Energy.—In all systems of common **battery** exchange the talking and switchboard signal energy is de**rived** from storage batteries which are charged from power-d**riven** dynamos in the usual fashion. This practice is superior **from**

FIGS. 168–169.—Motor-generator central battery ringing machines, the first designed to work from street power circuit ; the second, from storage battery at the exchange, both generating an alternating current. The machines are of the type supplied with the exchange battery plants of the Western Telephone Construction Co. The armatures are double wound, the commutator side being the motor, the collector ring side, the dynamo, the same field serving for both.

the fact that the fluctuations of current characteristic of dynamos, which are very annoying on telephonic circuits, are completely avoided without resort to transformers of any description. Again, since the source of power is seldom required for work at its full capacity at any one time, the operation of a dynamo would involve needless waste for most exchange systems. However, for signaling subscribers it is customary to use a power-driven alternating dynamo, such as is shown in Figs. 168–169, an alternating current being required to operate the subscriber's polarized ringer, while the direct current from the storage battery is required for the talking circuit.

Methods of Applying Central Energy.—In ringing from the central exchange to any subscriber's apparatus, the method of applying and transmitting the power need differ in no particular from the devices followed in the ordinary local battery systems previously described. How to centralize the talking energy is, however, a problem filled with many difficulties, since we must have a center source, a current in two directions, and an uninterrupted flow of energy from one subscriber's apparatus to any other with which his line may be joined at the switchboard. The most logical means of applying the current is to bridge the battery between the two strands of the plug circuit, so that one pole is joined to the sleeve strand and the other to the tip strand. By such an arrangement, as will be readily understood, a circuit will be made so soon as either plug of a given pair has been inserted in a spring jack; the current traveling from the battery on to the tip strand, thence through the jack contact spring to the subscriber's apparatus, back on the other limb of the line to the contact spring, through the sleeve, its strand, and to the battery again. To make the central energy operate like a local energy at each end of the line when both plugs are in jacks, or, in other words, to prevent the short-circuiting of the talking current at the central battery, is the difficulty first to be met and overcome. The various solutions of this situation constitute a big part of the several systems now in practical operation.

Connecting the Battery in Series.—One of the earliest applications of the central energy theory consisted in connecting the battery *in series ;* that is to say, attaching both its poles in one strand of the plug circuit, as, for example, the sleeve strand. The circuit was then made from the positive and back again to the negative pole, either when both plugs were inserted in the jacks of two subscribers, or else when the operator's talking set was bridged across the cords in communication with a calling subscriber. The receiver and transmitter at the subscribers' stations were also arranged *in series*, as in the ordinary local battery apparatus, the only difference being that the energy for operating the transmitter circuit was derived from the switch hook contact of the induction coil primary, as shown in Fig. 102, instead of from the galvanic cell.   While this system was perfectly practical, it would seem that a great disadvantage lay in the fact that the circuit, so energized, was not balanced, and that the speech-bearing current would be unevenly transmitted.

FIG. 170,—" Dynamotor," charging machine for charging the storage battery.   The motor, worked from street power circuit, is attached on the same shaft with the dynamo.

Methods of Balancing the Circuit.—As a consequence of its shortcomings this system of battery connection was abandoned for that at present adopted—the method of bridging the battery between the plug strands.   By this construction, as we have seen, it is perfectly practicable to transmit a current in two directions—from central to both subscribers' stations and back again—but the theoretical difficulty is to ensure a through current from one station to the other, and prevent short circuiting at the bridged battery.   To accomplish this result three typical devices have been adopted: 1. To interpose an impedance coil between each pole of the battery and the strand to which it is connected, as in the system devised by John S. Stone. 2. To interpose a divided repeating coil between each pole of the battery and the strand to which it is connected, each strand being thus divided, and the double windings of the coils being joined at the pole of the battery, as in the system devised 'r Hammond V. Hayes. 3. To connect the center points of

both strands of the plug circuit with the winding of an impedance coil, whose center point in turn is connected to one pole
of a grounded battery, as in the system based on the conceptions
of J. J. Carty and W. W. Dean. Most of the more recent
central energy devices are based on these broad principles or
consist in some variation of the systems about to be described.

The Stone Central Battery System.—By interposing
impedance coils between the battery poles and the connections
on the plug circuit strands two ends are achieved; the direct
current flows from the battery and back again with ease, the low
ohmic resistance of the coils offering no bar whatever; the
rapidly alternating telephonic current is effectually confined to
its line on the plug strands by the magnetic action of the coils,
and short-circuiting through the battery is completely prevented.
Thus the battery current, dividing on the plug strand and
flowing in two directions to both stations and back again, is
able, as it is modified by the variant pressure in the transmitters,
to convey the telephonic impulses from one end of the circuit
to the other. In connecting a number of plug circuits to one
battery the method is to connect both strands of all plugs with
the impedance coils and join them *in multiple* to the battery
poles. That is all sleeve strands are joined through the sleeve
coils to the positive pole of the battery, and all tip strands,
through the tip coils, to the negative pole. The subscribers'
apparatus in this system are constructed with the receiver and
transmitter connected *in series* to the line wire, as in the ordinary
exchange instruments, the current for the transmitter being
usually derived as in the *series* battery system described above.

The Hayes Central Battery System.—In the Hayes system
the battery is bridged or wired *in multiple* between the plug
strands, as in the one just explained, the principal difference
being that each pole is connected to the cord circuit through a
divided repeating coil instead of through an ordinary magnetic
impedance. In other words, the current is passed in two
directions through a transformer or induction coil resistance,

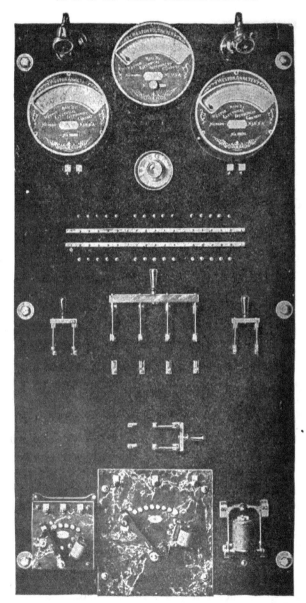

FIG. 171.—Power switchboard for a central energy exchange.   This is a slab of slate upon which are mounted the various devices shown for controlling and measuring the strength and amount of the current, also rheostats for controlling the charging current from the dynamo and the current from the storage battery, and cut-out switches.

the end attained being to obtain the same kind of an inductive effect between the two extremes of the line as is obtained between the primary and secondary circuits of the ordinary telephonic transmitter. In this way, while the direct current from the battery emerges from the positive pole, travels along both windings of the repeating coil to both stations and back again, the continuity of the alternating telephonic current from one end of the line to the other is maintained by inductive action. Such an arrangement seems to possess points of superiority, since, while the plan of passing the battery current through magnetic retardation coils effects the end of holding the telephonic current on the line, the use of a transformer utilizes the central *leak* to supply an inductive reproduction of the transmitted waves.

Of Repeating Coils.—Repeating coils are used for a variety of functions in telephony, being particularly efficient in balancing a circuit, as, between a grounded and a full metallic line. In the Hayes common battery system they differ from ordinary induction coils in that both windings are of the same size of wire and of the same length instead of having a primary of heavy wire with few turns and a secondary or thinner wire with many turns. Like the induction coils already described, all repeating coils produce an effect of accelerated current intensity according to the number of turns of the secondary winding; the ratio, when other conditions are equal, is just about equivalent to that between the turns of the two windings. Telephonic repeating coils are usually constructed on the principle of the Faraday ring, shown in Fig. 172, which differs from the ordinary induction coil or other transformer depending on inductive action of one winding upon another, such as is shown in Fig. 173, in the fact that the core is itself a continuous coil instead of a bundle of short wire lengths or a solid iron bar. The primary and secondary windings, instead of being superposed, as in the induction coil, are made on either side of the ring, as shown in the figure. If they are so superposed, as in the induction coil, the core winding is passed through the eye

of the coil and completely around so as to envelope the primary
and secondary on all sides, as shown in a later figure.

In Fig. 172 of the Faraday ring, taken as a typical transformer, suppose *P* to be the winding, one end of which is
attached to the sleeve strand of the answering plug in the
Hayes system, and *S* the winding, one end of which is attached
to the sleeve strand of the calling plug. The opposite ends of
both helices are joined and attached to the positive pole of the

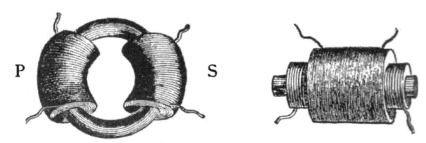

Figs. 172–173.—Illustrating two types of transformer. The first is the " Faraday
ring," a coil of iron wire on which two separate helices are wound. This is the model for
the standard repeating coils. The second is an ordinary double wound transformer with
" laminated " core, like the common induction coils.

battery.   The same is true of the repeating coil used to connect
the tip strands of both plugs with the opposite pole of the
battery.   Thus the current emerging from the battery is divided,
one-half passing to the line and apparatus of the calling subscriber through the answering plug, and the other passing to
the line and apparatus of the called subscriber through the
calling plug of the given pair; returning in similar fashion
over the test wire of both subscribers, through the tip contacts,
strands and repeating coil, to the opposite pole of the battery.
The coils used in this system are usually wound to a resistance of
about five ohms, the double windings being equal, both as to
size of wire used and the number of turns, thus balancing the
line.   It follows, therefore, that any variation in the total line
resistance produced in the transmitters will be evenly repeated
through the full length of the circuit, and that a balanced
inductive effect will take place between the two helices of both

coils, either helix acting as primary or secondary. The usual practice is to wind both coils, four helices, on the same core.

The Dean-Carty Common Battery Arrangement.—A third system of transmitting energy from a central battery to two connected stations consists, as we have seen above, in bridging a long-wound impedance coil of low ohmic resistance between the strands of the plug circuit, one end being connected to the sleeve strand, the other to the tip strand, and attaching one pole of a grounded battery to its center point. By this arrangement the battery current moves from the center point of the winding in both directions, running along both strands of both plugs, on both line wires, to both stations, and finding its return path from both subscribers' apparatus through the ground to the opposite pole of the battery. The impedance of the coil prevents the telephonic current from being shunted from the metallic line on to the grounded battery circuit, and while maintaining a balance throughout, is particularly efficient in insuring through conduction of the transmitter impulses.

Of Impedance or Retardation Coils.—An impedance or retardation coil consists of an iron core wound with a number of turns of insulated wire, and in this particular is comparable to an electro-magnet, an induction coil or one of the type of transformers just described. Its principal function in telephony is to bar or impede the alternating speech-bearing current so as to keep it on its destined line and off of a bridge between the limbs, or from an associated circuit, as in the test arrangement of a multiple switchboard or the battery wires of a central energy system. This result is to be achieved not by a high ohmic or line resistance in the windings of the coil, but by the reactance of the magnetized core. Thus, although the term "impedance" properly designates any form of check or obstacle met by the electrical current, including true, ohmic resistance, it is most generally used as a synonym for the effects of magnetic retardation on alternating currents. In order to attain this effect to the best advantage it is essential that the core be made of sufficient mass to allow of the greatest desired degree of magnetic reactance, and of such length as to permit a large number of turns of the winding wire. With a core of the same mass and a wire of the same size or weight, the efficiency of two given retardation coils is usually to be measured by the length of the core and the number of turns. Thus for the winding a wire of comparatively low resistance may be employed, and no particular check offered to the free transmission of a direct current, whose strength and induced magnetic properties are moderately constant. An alternating current is impeded in propor-

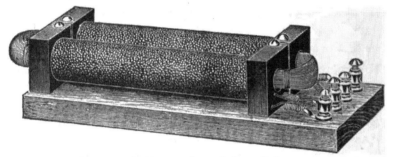

FIG. 174.—Couch & Seeley's double-wound retardation coil, for use on central energy exchange circuits, and common battery private telephone systems.

tion to the variations in these particulars. For this reason the high-wound bridging bells, such as we have noted in connection with the central energy systems already described, will bar the telephonic current, which is immensely irregular in phase and frequency, while the magnets may be readily energized by the regularly alternating current from the exchange ringer generator. In measuring the impedance of such instruments in ohms it is, of course, intended to express the sum total of all checks to the telephonic current, including the "false" as well "true," or ohmic, resistance.

Circuit Arrangements in the Dean System.—The impedance coils used in the Dean central energy system are constructed with a flattened ring core, on the principle of the Faraday ring shown in Fig. 172, except that there is but one helix instead of two. This arrangement is adopted in order to give a complete magnetic circulation with a corresponding increase in the retardation, although the ohmic resistance to the battery current is very low. The winding of the cores is double; that is, two wires are wound side by side to the full length of the helix, as is the case with the secondary of the induction coil described in connection with Fig. 93. One end of one of them is attached to the sleeve strand of the plug circuit, one end of the other to the tip strand, and the opposite ends of both are brought back, joined together and attached to the battery pole. Thus the current emerging from the battery is divided into two parts, both of which move through the entire length of the helix of the impedance, and thence over both limbs of the line circuit *in multiple* to the subscribers' stations.

The Subscriber's Apparatus Circuits.—Just as there is an impedance coil bridged between the plug strands at the exchange, so there is one of approximately the same resistance and retardation bridged between the two sides of the line circuit at the entrance of the subscriber's apparatus. At the exchange the current flows from a grounded battery to both ends of the double winding; at the subscriber's apparatus it flows *into* the coil *from* both ends, joining into one current again at the center point of the double winding. Beyond the points where the coil is bridged between the line and test wires, the receiver and the secondary of the induction coil are strung *in series*, the circuit being completed between them. The transmitter and primary winding of the induction coil are placed in a permanently closed circuit, and at the center points on either side are attached the bridge leading from the center point of the retardation coil, on the one hand, and the ground terminal of the battery circuit, with an interposed resistance coil of high ohmage, on the other. Moreover, the primary winding of the induction coil is double, the two halves being wound in opposite directions, so as to ensure an additional inductive effect on the secondary; an increase of resistance in the one equaling a decrease in the other. The two windings then join together so as to allow the current to go to ground through the resistance

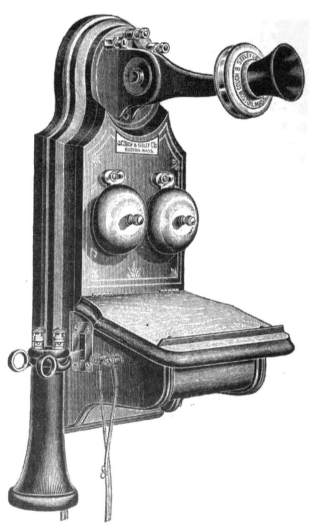

FIG. 174a.—Central Energy Telephone Apparatus. The back plate of this apparatus is
a hollow iron casting, sufficiently large to contain a condenser, which, on the common
battery system of the manufacturers, is connected *in series* between the bridge containing
the polarized ringer on the one hand and one-half of the transmitter and primary circuit,
on the other. Energy for ringing the signal bells comes in over the telephone circuit wires.

coil just mentioned. The bell magnets, which are high wound, are in circuit so long as the switch hook is down, being grounded, so as to respond to the impulses sent along the line from a grounded magneto-generator or dynamo at "central." The removal of the receiver from the hook completes an auxiliary circuit from the exchange grounded battery through the helix of the rising visual signal, this circuit being broken at the jack by the insertion of a plug.

Improved Systems.—Other applications of the general principle of imparting energy through the two strands of the plug circuit differ from those described above mostly in the elaborations consequent upon the management of supervisory signals to keep the operator apprised of the condition of the talking circuit, and such additional devices as are required in the operation of multiple switchboards, where, as may be understood, there must be suitable arrangements for cutting out the line signal circuit at every multiple jack. In general, this end is attained by the order of devices described in connection with the use of incandescent lamps as calling signals. The best practice, in the estimation of many of the inventors who have attacked the problem, is to pass the central battery current through impedance coils, at least one system introducing the variation on those described of bridging a grounded battery circuit between the plug strands, the grounded battery being attached to one strand and the opposite grounded terminal to the other, both through impedance coils.

Other Methods of Imparting Energy.—Some inventors have attacked the problem of imparting energy to the transmitter circuit from a different starting point. Instead of attaching a battery between the strands of the switchboard plugs, they connect the primary circuit of the subscriber's apparatus with a power line, additional to the telephonic line, to and from the central exchange. One of the simplest, and, in many respects, one of the best methods under this general head is that described in connection with lamp signal circuits, whereby a constant current battery continually charges a storage cell at the subscriber's apparatus, so long as the hook is down, the high resistance of the ringer magnets preventing it from operating the switchboard signal. In this system, as we have seen, when the storage cell fails of a sufficient charge to operate the talking circuit, sufficient current may be derived direct from a bridge connection of the primary winding of the induction coil with the line wire to give the desired effect. In this case the telephonic circuit wires are used for conducting the power current until the circuits are changed by raising the switch hook.

Systems Operating on Separate Circuits.—Many of the later methods of applying centralized energy direct to the station transmitters operate by employing a separate power circuit, making four wires in all to and from each subscriber's apparatus. Two of these wires are the line connections of the talking current; the other two bridged connections to a common, or street, power circuit.

Dynamo Circuit and Thermopile.—One brilliant device, invented by Mr. W. W. Dean, is to include the subscriber's transmitter and primary winding of the induction coil in a permanently closed circuit with a thermopile. The junctions between the plates of this pile are made, first on one side, then on the other, so that each *alternate* junction is approached to a resistance coil, each other *alternate* junction being in the opposite direction. The resistance coil is included in one side of the separate power circuit, so that when current is passed through it, it heats and generates current in the thermopile sufficient to operate the transmitter. The power circuit, like the talking circuit, is made by the rising of the switch-hook, a contact piece, insulated from the shank of the hook, being provided for that purpose, the insulation being intended to prevent the dynamo current from being shunted on to the line wires.

Storage Cells and Condensers.—Mr. J. J. Carty has devised a method for energizing the transmitters direct by attaching the two sides of the *power circuit in multiple* to several low resistance storage cells at the exchange. This arrangement

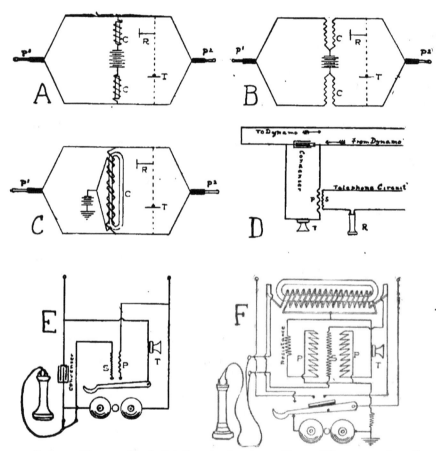

FIG. 175.—Diagrams illnstrating the several systems of transmitting energy from the exchange. *A*, the Stone system, plug circuit; *P*₁, *P*₂ plugs, *C, C*, retardation coils, *R*, operator's receiver, *I*, transmitter. *B*, Hayes system, plug circuit; *C, C*, repeating coils. *C*, the Dean system, plug circuit. *C*, the retardation coil. *D*, feed circuit with condenser *in series;* *P*, primary. *S* secondary, *I*, transmitter, *R*, receiver. *E*, apparatus wiring of the Hayes system. *F*, apparatus wiring of the Dean system.

may be understood by reference to Fig. 47. The transmitter circuit of each subscriber's station is then connected *in multiple* to the power circuit, which arrangement may be understood by reference to Fig. 48, *L, L, L, L,* in this case representing so many subscribers' transmitters with the induction coil primary on bridges. The danger of cross-talk in this arrangement is obviated by making the batteries of low resistance and the power conductors as short and heavy as possible in comparison to the transmitter and primary circuits, so that the fluctuations of current produced in one apparatus will return direct to the source rather than leak across any of the other bridges. This arrangement renders all drops in potential inappreciable at the battery terminals.

Series Feed Circuits.—There are two noteworthy systems of imparting energy for the transmitter from a *series* connection. In both of these the circuit of the transmitter and primary winding forms a *shunt circuit*, its two terminals being connected to the wire, as may be understood by reference to Fig. 46, each *L, L, L,* in this instance, representing subscribers' transmitter circuits through which the current passes *in series*. Across the shunt circuits, however, and between the two terminals just mentioned is bridged, on the first system a storage cell, on the second a condenser, each also *in series* with the power wire. Consequently when the transmitters are operated by the voice, sufficient additional energy may be derived from the storage cells, in the first instance, for distinct transmission. In the second instance, the current fluctuations produced in the transmitter vary the potential of the charge in the condenser, around which it is shunted, and gives the full effect of an alternating current on the primary winding of the induction coil. Such alternation is then said to be "superimposed" upon the direct dynamo current always flowing through the feed circuit and transmitters. In the latter system an impedance coil is interposed between each pole of the dynamo and the terminal of the line power circuit.

CHAPTER SEVENTEEN.

PARTY LINES AND SELECTIVE SIGNALS.

Party Lines or Common Circuits.—In all of the systems we have considered heretofore, the apparatus of each subscriber is connected to the central exchange by its own circuit.
There is, however, another method frequently adopted in
systems where there is but little business in proportion to the
length of the line, and it is, briefly, the connection of a number
of subscribers' station apparatus on one circuit, so that all have
a common drop and jack on the exchange switchboard. This
is called the "party line" or common circuit system, and it is
also available for small private telephone systems, connecting a
number of neighboring houses with one another or with stores
or places of business.

The Series and Bridging Systems.—In the party line
system of telephones, as in the case of electric light, power and
other circuits, there are two general methods of connecting the
different stations. They are connected either in series or in
multiple. The latter method is better known in telephony, in
its simplest application, as the "bridging bell" system.

The Series Method of Wiring.—The series method of
connecting a number of telephones upon a party line is shown
in Fig. 176. Here, as may be seen, one side of the line extends
from one apparatus to the next, and passing through the coil of
the signal bell in each—the bell circuits being closed so long
as the receiver is hung on the switch hook, as we have seen—
emerges at the other binding post at the top of the generator
box, whence it goes on to the next station. The other side of
the line extends from the right-hand binding post of the right-
hand instrument back to the left-hand binding post of the left-

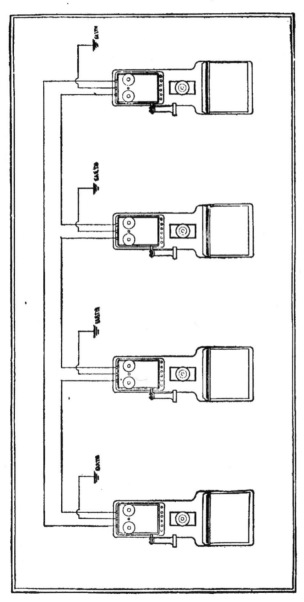

FIG. 176.—Diagram of a full metallic series telephone party line circuit, showing four station apparatus connected to the line wire.

hand instrument. In grounded lines the ground connection is made at these two extremes only.

The Wiring of a Series Apparatus.—The wiring of a series circuit party line station telephone apparatus is substantially the same as that of the ordinary exchange circuit apparatus, as may be seen by reference to Fig. 177. As in the latter instrument, the hook makes two contacts when the receiver is removed, thus making the talking circuit, and but one when the hook is down, thus making the calling circuit.

In calling in a series system a special code is used, by which each telephoner rings out an agreed signal to call each other one. Such signals are generally mere indications of the numbers of the other operators' apparatus, and may be given by turning the crank of the generator with alternate pauses instead of continuously, as in the case of the subscriber who desires to call up the central station, as we have seen.

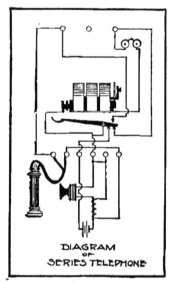

DIAGRAM
OF
SERIES TELEPHONE

FIG. 177.

Objections to the Series System.—As may be readily understood, the principal objection to the series system of party line telephones is that each conversation must be carried on by a current that passes through the call bell coils of every other apparatus on the line. Such an arrangement is bound to weaken the current in proportion to the number of apparatus connected, and for this reason the bell magnets·have usually been wound to a much lower resistance than in the ordinary type of telephone; sometimes as low as 80 ohms, just about the average resistance for receiver coils, for each double-pole winding. The telephonic current, while it may pass through a number of such coils and still remain at hearing intensity, is not sufficiently strong to energize the bell magnets. Thus, although the high potency currents from the magneto-generators can readily ring through the combined resistance of fifty instruments (4,000 ohms), the talking current at that figure becomes so faint, through the combined effects of line resistance, impedence, induction and general leakage, that it is extremely unsatisfactory, not to say impracticable. Such troubles may be partially overcome by frequently transposing the line; for example, connecting Station I to Station IV, and Station II to Station V, etc., instead of connecting in succession, as in the cut, or else by connecting stations alternately to either side of the line. Even with this arrangement balance cannot be maintained, and the talking qualities are only partially improved.

Magneto-generators for Series Circuits.—As it is often necessary to ring through as high a resistance as 4,000 ohms, combined with line resistance, on such a line as is mentioned above, it follows that the generators of a series telephone circuit must be of high electromotive force, sometimes reaching so great a strength as to ring through 15,000 ohms, or about four times the power of the ordinary exchange instrument.

The Bridging Bell System.—With a view to overcoming the serious difficulties thus mentioned, another system of stringing a telephone party line has been introduced within a few years. On this plan the several telephone apparatus are strung in parallel or multiple, as shown in Fig. 178, which shows the arrangement for a metallic circuit line having four stations. In a grounded circuit the ground connections are made direct from the right-hand top binding post of each instrument, which in this diagram is shown to be connected to the return wire. The ground connection shown at the center binding post of each apparatus is for protection against lightning and electrical disturbances, or else, as in some systems, to afford the necessary ground connection for some scheme of selective signaling.

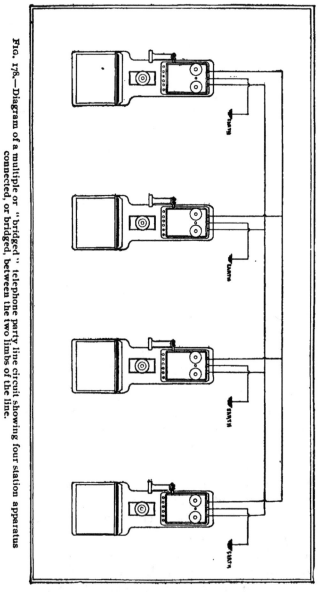

FIG. 178.—Diagram of a multiple or "bridged" telephone party line circuit showing four station apparatus connected, or bridged, between the two limbs of the line.

The Carty Bridging System.—The bridging system which was originated and patented by Mr. J. J. Carty, of the Metropolitan Telephone Co., New York, is, in many respects, the opposite of the series, and certainly overcomes most of its gravest faults. The first advantage is. found in the fact that the telephone current does not pass through the coils of the several bell magnets, as in the former case, and these are wound to a very high resistance, a large number of turns and long bobbins, so as to effectually prevent the shunting or short circuiting of the current before it reaches the apparatus to which it is destined. In short it may be said that the whole theory of the bridging telephone system may be summed in the principle that a current will always follow the line of least resistance or of greatest ease in making a circuit. This principle is applied in the interior arrangement of the apparatus as well as in the stringing of the line.

Circuits of a Bridged Telephone.—Fig. 179 shows the circuits of a bridged telephone apparatus. As may be seen, the rising of the hook on the removal of the receiver, while it makes the circuit of the transmitter and receiver, does not break that of the bell and generator, which are left permanently bridged across the talking circuit. For this reason the switch lever makes no lower contacts. Two possible paths are thus presented to the telephone current: (*a*) that leading through the bell magnets to line, or (*b*) that leading direct to line. It chooses the latter path, as being the line of least resistance. Similarly a current on the line will enter an apparatus in which the telephone circuit is made, and will avoid the path leading through the coils of the bell magnets, which are bridged across.

Wiring of a Bridged Line.—As may be seen from a brief examination of Fig. 178, a complete circuit may be formed between any two apparatus in the system without reference to any of the others. Thus, if the telephoner at the second station is conversing with the one at the third, the current passes out from the right-hand binding post of his apparatus, through the line wire, into the apparatus of Number 3, through the right-hand binding post, through his talking circuit, out of his apparatus by the left-hand binding post, to the return side of the line, and thence back to apparatus Number 2, through the left-hand binding post, thus completing the circuit. The high resistance and retardation of the bell magnets of the first and fourth apparatus, which remain bridged across the line, forming so many shunt circuits for the current from the magneto generator of any one of them, which, when operated, rings all the bells in the system, effectually bar the talking current on the same principle as that applied when the coil of a switchboard clearing-out drop is left bridged across the line of two subscribers. The resistance is so high that the talking current has no tendency to enter the coil, and remains on the line between the two apparatus whose talking circuit is made.

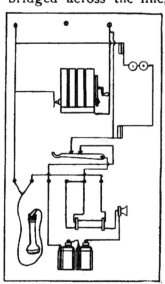

FIG. 179.

High Impedance of the Bell Magnet.—For the purpose of barring the talking current, the coils of the bell magnets are wound with a greater number of turns of wire than ordinarily, generally to a resistance of about 1,000 ohms, although a heavier quality of wire is usually employed than in ordinary coils. This latter practice has been adopted for the purpose of winding a longer core, thus obtaining more iron in the magnet and increasing the magnetic retardation. Coils so constructed, while readily operated by the low frequency currents from magneto generators, are effectual blockades against the high frequency currents of the talking circuit. As has been calculated, the impedance of the bell magnet coils is about four times its ohmic, or true, resistance during the passage of the telephonic current. The question of high retardation is of much greater importance than mere resistance in the coil wire. Thus the use of finer wire, or some highly resistant metal, such as German silver, while giving the same or a higher degree of resistance, is not so good a practice as using a coarser wire. The latter practice secures the required length for the magnet cores without undue increase in the resistance, while the former would only impede the path of any but the most powerful type of magneto-generator current, if there should be an attempt to give the usual length to the magnet cores.

Fig. 180.—Interior of the generator box of a bridging telephone apparatus, showing the long-wound ringer magnets. Compare with Figs. 3, 97, 98.

The Generators of Bridged Telephones.—The generators of bridging telephones are constructed with the express view of securing a large quantity of current, a high amperage, so as to ring the bells of the most remote apparatus on the line. They must also be constructed to give a high voltage to work through the combined resistance of all the coils. The armatures are usually wound to a resistance of about 350 ohms, with No. 33 B. & S. wire, which gives about the right current strength for ordinary bridged lines. Some manufacturers have sought to strengthen the current producing capacity of the generators by increasing the number of the magnets to four or five, thus giving a longer field and allowing the winding of more wire on the armature—giving a greater number of "ampere turns"— with a consequent augmented E M F. When, however, the bridging circuit is connected to a central exchange, and some form of selective signaling is there installed for calling up the desired station, the generators of the separate apparatus are wound to furnish a low E M F, which will be effectually barred by the bell magnets, while amply sufficient to operate the switchboard drop, or signal, as in ordinary single-station exchange circuits. Another method of accomplishing the same result is to wind the station magnetos so as to generate direct instead of alternating currents, which, while readily operating the bridged drop of the exchange switchboard, cannot energize the polarized bell magnets wound for alternating currents.

Induction Coils Low Wound.—On the other hand, it is very essential in any apparatus intended for a bridging circuit that the induction coil secondary should be wound to as low a resistance as is consistent with good transmission. Otherwise, and particularly when the resistance of the secondary winding is as high as 500 to 800 ohms, it happens that the talking current will seek a path through the bell magnet coils as quickly as through the induction coil secondary and the receiver.

Independent Bridging Systems.—The Carty patent, covering a system of apparatus wiring best adapted for use in bridging circuits, is the one under which most manufacturers of such telephones operate. As, however, it does not cover the method of stringing an inter-communicating line in multiple, others have devised systems of their own which accomplish the same ends without infringing. Fig. 181 shows a circuit diagram of the Ericsson bridging or multiple-wiring system. As may be seen at a glance, it differs considerably from the Carty wiring, already shown. When the hook is in the lowered position under the weight of the receiver, the circuit of the bell is made at the end contact of the switch hook lever. When it is allowed to rise, on the removal of the receiver, it makes two contacts, as shown, thus closing the talking circuit. The principal differences between the Carty and Ericsson wiring are, therefore: (*a*) the bell magnets of the communicating apparatus are not bridged across the line when the talking circuit is made, and

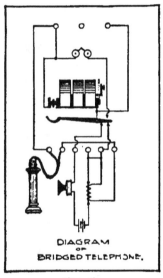

DIAGRAM
OF
BRIDGED TELEPHONE.

FIG. 181.

the neglect to return the receiver to the hook at the close of the conversation would prevent the ringing of the signal; (*b*) the switch lever has both upper and lower contacts, as in all other types of apparatus; (*c*) the bell coils of the two communicating telephones being cut out of line, all "leakage" of current is limited to the apparatus not so connected, and all unnecessary resistance is dispensed with. The Ericsson Company claims this latter feature as the greatest advantage of the system. They, however, follow the general practice of high winding and long cores for the bell magnets, frequently using a resistance as high as 1,600 ohms, which still further decreases the tendency of the telephone current to "leak," and, as they claim, secures the best possible talking service.

Selective Signals.—The greatest possible disadvantage of all party line telephone systems is that the act of "ringing up" one station, even by an agreed code signal, rings the bells at every other, thus constantly annoying every party on the line, making possible frequent mistakes and consequent confusion, and, worst of all, allowing every conversation to be "listened in" by any meddlesome fellow with an apparatus in his house or place of business. Such facts have led numerous inventors to devise schemes for "selective signaling," as it is called, by which a signal intended for any particular apparatus will be rung by its bell only; and also methods of "locking out," whereby the act of making the telephone circuit between any two apparatus energizes conveniently placed relays in every other, thus cutting the other apparatus out of line. While it is not within the province of this book to enter into the details of the numerous ingenious and complicated devices for accomplishing the end of selective signals, it will be possible to indicate briefly the general varieties of such machines. These fall under three general heads—those working on the "step-by-step" principle; those working by the use of harmonic reeds; those working on the principle of varying "strength· and polarity" of bell magnets and calling generators.

The Step-by-step Principle.—The first of these, the step-by-step signal system, is at once the most practical and most familiar in its operation. It is constructed on the same general principle as is applied in the dial, or needle telegraph, and in the ordinary stock quotation and news "tickers." This principle consists, briefly, in transmitting a current, controlled by a regularly graduated make-and-break device, which operates relays at the various stations on the line. The transmitting instrument is generally a dial divided into a number of points, each corresponding, as in the "stock ticker," to some particular letter or figure. Each such point, moreover, is a "make," while the spaces between them are "breaks." By moving the key, pivoted at the center of the dial, over a given space, say from *A* to *M*, one makes and breaks the circuit as many times as there are points and spaces between, thus sending precisely the same number of *distinct* impulses along the line as there are make points passed over before reaching the desired point, say *M*. These several impulses operate the relays at the several stations exactly the same number of times, and by virtue of a pivoted lever attached to the armature of each, works a ratchet wheel or escapement, such as is used with clock pendulums, so as to operate a needle on a lettered dial or a wheel bearing printing types on its periphery. On bringing the above-mentioned transmitting key to the point desired, say *M*, the needle of the dial rests, or the mechanism of the "ticker" causes that particular letter to be impressed on the paper strip, which then moves one point ahead for another impression.

Step-by-step Signals.—The application of this principle to telephone party line circuits, while, seemingly, requiring much more complicated devices than those described, may be briefly stated as follows: The "make" points on the transmitting instrument correspond, not to letters or figures, but each to some particular apparatus in the circuit. Thus, if there are eight apparatus in the line, the first corresponds to the first point, the second to the second, and so on to the eighth. Moreover, the toothed wheels at each station are so arranged by insulation of every point except the one corresponding to its proper number, say, one, two, three and so on, that its call bells may be operated only when that point has been brought into position. We may understand this by an illustration. Suppose that a number of printing telegraph instruments be arranged in a circuit. If from each of the type wheels, save one, we cut the letter *M*, it is obvious that by that one only w.ll the letter *M* be printed. On the same principle the bell of apparatus No. 1 can be operated only when the relay and ratchet has brought the signal controller to point 1; similarly with all other apparatus. If, therefore, a telephoner wishes to call up Station 3 he moves his circuit-controlling key to the 3-point, and then sends out the ringing current, with the effect of calling Station 3 and no other on the line. The other call bells are effectually · locked out by virtue of the insulated contacts at all points save that corresponding to their own numbers.

The Transmission of Harmonic Waves.—The harmonic system of selective signaling is based upon the same principles which we have already seen applied in Prof. Bell's first experimental telephone and other contrivances for the transmission of musical sounds. The calling current consists of waves of different length and frequency, according as it is started by the

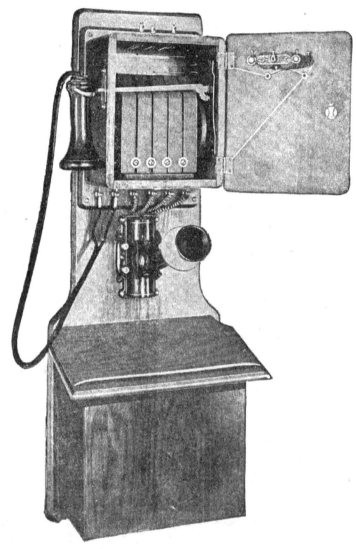

FIG. 182.—**Western Bridging Bell Party Line Telephone Apparatus, equipped with a five-magnet generator.**

vibrations of a circuit-cutting reed or rod at the calling station. Furthermore, the making of the calling circuit at each apparatus depends upon the reception of a current of the right degree of frequency and wave length to affect a circuit-cutting reed or rod, similar to the one vibrated at the calling station. Such receiving reed or rod at each instrument differs in these respects from the one at every other station. The idea may be briefly expressed by saying that each apparatus has its own note—the rate of vibration of its reed—which must be sounded at the calling station before the current can ring the bell of that apparatus. Thus if a party line circuit be operated by harmonic signals, and be not connected to a central station, each apparatus must be equipped with its own receiving reed and also with some device for shortening and lengthening the vibrations of the calling reed mechanism. Most selective signal party line circuits, however, are constructed with reference to the connection of the line to a central office, where suitable signals may be operated by the calling current from any one of the apparatus on the circuit, and the complicated calling mechanism may be operated to the best advantage.

The Pendulum Signal.—One of the simplest and most efficient signaling devices of the harmonic, or wave length, description is the Stephen vibrating pendulum system, which has been used to some extent in England. It may be briefly described as follows: In each station is a pendulum of fixed length, different at each station, and having a proportionate rate of vibration. Attached above its pivot is a fork or hook standing at right angles to the length of the pendulum rod. The pendulum rod itself carries the armature of an electro-magnet, and is set vibrating by any calling current that may enter the apparatus. The fork or hook above mentioned is of such a length that, when the pendulum is actuated by a current of the right wave length and frequency, and vibrates accord-'ngly, it will release a hooked lever arranged at the right terval and thus make the circuit of the call bells by breaking ; normal or resting connections. This normal position is restored by a double-pole relay, which acts on another lever, so arranged as to renew the first connections under impulse of a current sent from "central," when the conversation is concluded. The principal point of the selective signaling in this apparatus is the length of the fork or hook at the upper end of the pendulum rod, which, with the relative position of the lever to be actuated by it, differs in every station apparatus on the circuit.

The Metronome Transmitter.—The apparatus used to send a call from "central," or from any calling station, consists of an instrument resembling the *metronome* used by musicians for beating out the time required for any musical composition—a rod which, moved by clockwork, vibrates slowly or rapidly as a weight is slid up or down to one or another of the graduated positions in its length, thus shortening or lengthening the rate of vibrations. With such an instrument to control the undulations of the current—being used as a rheotome—the selective signal may be readily transmitted according to the position of the weight on the indicated marks corresponding to the rate of vibration for one or another station· apparatus. Such a method of signaling, while simple in theory and perfectly practical, is open to grave objections on the ground of the inevitable delay in making connections on the line.

Signaling by Current Strength or Polarity.—The third head under which party line selective signal apparatus are grouped is that of varying the strength and polarity of the generators and bell magnets at the several stations. A number of highly elaborate systems have been devised to apply this principle to party lines of from three to eight stations on a single circuit, the calling impulses in every case being sent from a central office switchboard, where suitable ringing switches are arranged. In all these systems the feature of bell magnets, so polarized as to operate in obedience to a current coming from one direction and remain irresponsive to one coming from another, is prominent, the principal points of divergence being

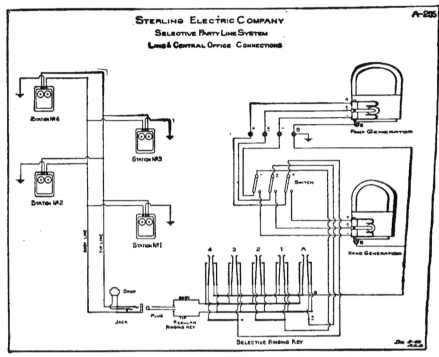

FIG. 183.—Diagram of the four-station polarity signal system of the Sterling Electric Co.

Stations are to be connected to line as shown on diagram, that is, the right hand post on magneto bell on stations one and three are to be connected to the tip side of line, and the left hand post to the body side of line. At stations two and four, the right-hand post is connected to the body side of line, and the left-hand post to the tip side of line. The middle post at all stations is to be connected to the ground. The generators are supplied with commutators, so arranged that they will furnish both the alternating and straight currents with a grounding point, which must be connected with a good ground.

A selective key is furnished, connected in such a manner that any desired current can be thrown from the generator to any regular ringing and listening key on the board. If hand generator only is used for operating this system, the positive and negative brushes of the commutator, which are indicated on the diagram by the signs (+) plus and (—) minus, respectively, are wired directly to the two short bus bars on the selective key, which are also marked with a (+) plus and (—) minus sign to correspond. The alternating spring or brush which is designated by the (+) plus and (—) minus signs in combination, is carried to the isolated point on the selective key correspondingly indicated. Where both power and hand generator are to be equipped with the party line commutator, the connections must be complete, as shown on diagram, going through a three lever switch. The power generators must invariably be run in the direction indicated by arrow stamped on the face of the pulley. A reversal of this direction will result in transposition of the numbers of the stations responding to calls owing to the fact that the current will be reversed.

In the operation of the system in order to call station No. 1 on any party line, the button marked No. 1 on the selective key must first be depressed, the plug inserted in the jack in the usual manner, and the ringing done on the regular ringing and listening key as usual. Pressing the button No. 1 merely throws the current required to operate station No. 1 to that pair of plugs. If No. 2 is wanted, No. 2 button is depressed, and so on. When generator is to be used for ordinary exchange work, calling only for alternating current, the button on selective key marked "A" must be first depressed. Where the subscriber on one of these party lines wishes to call the central office, he must first take the receiver from the hook before ringing, as otherwise it would be impossible to ring down the drop at the central office. This will also give him an opportunity to ascertain whether the line is in use before calling for connection.

in the means employed to apply this scheme more readily, or to vary the number of possible combinations of circuits and polarized magnets.

A Two-limb, Three-circuit Signal.—Most of the later applications of the principle contain some variation of the idea first introduced with the Sabin and Hampton three-station system of making three distinct circuits with the two limbs of a single metallic line, consisting of a line wire and a return wire, as must be now understood by the reader. This idea is, briefly, to place the call bell magnet of the first station in a "bridge" between the line and return wires; to attach one side of the bell circuit of the second station to the line wire and the other to ground; and to attach one side of the third station to the return wire and the other to ground. To operate the signals on this system it is necessary that there be three separate generators at the central station, each one being connected to the wires in precisely the same fashion as are the call bell magnets. Thus the current from one goes out on the line wire and returns on the return wire; that from the second goes out on the line wire and returns by earth connection; that from the third goes out on the return wire and returns by earth connection. As will be obvious on reflection, a current sent from any one of the three generators completes its circuit through the instrument it is intended to affect, and through no other.

Six Signal Circuits to Two Wires.—A further elaboration of this system may be obtained by stringing six bells instead of three, two on every circuit just described, by alternating the polarity of the bells and generators. Thus there may be two call bell magnets at separate stations, each bridged between the line wire and the return wire, but one of them must be so polarized as to respond only when a positive current comes on the line wire and a negative on the return wire; the other, so as to respond only when a negative current comes on the line wire and a positive on the return wire. Similarly, also, the grounded circuits on each limb may be duplicated by making one magnet of each pair to respond only when a positive current comes from earth and a negative from the wire. To accomplish the result of selective signaling in both cases it is necessary to have either six generators, properly polarized, or some method of changing the circuit connections from one side of the line to the other, so as to send a positive or a negative current by the desired route. Thus a simple three-circuit arrangement may become a true polarity method of signaling. Such an arrangement might operate to advantage on a short line.

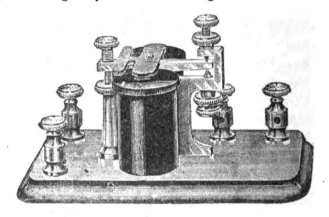

FIG. 184.—A common form of telegraphic relay. A telegraphic "sounder".

The Principle of Quadruplex Telegraphy.—In addition to this feature some systems also embody the idea which is basal in quadruplex telegraphy—the introduction of relays which respond only to current of given strength and are indifferent to its polarity. In telegraphy this principle is applied by having a "polarized relay" and a "neutral" or current strength relay at every station, each relay being operated by a separate key in the transmitting office. Thus, as may be understood on reflection, this arrangement admits not only of selective signaling, but also of four messages being transmitted at one time over one grounded circuit, two in one direction and two in the other, the strength of the current and its polarity being the important considerations in each case.

The Merits of Selective Signal Devices.—As regards the matter of selective signals in general, we can do no better than to quote from Herbert Laws Webb's excellent "Telephone Handbook." He says: "Many inventors have turned their attention to this problem, and T. D. Lockwood announces in a paper before the American Institute of Electrical Engineers, read in 1892, that between January, 1879, and December, 1891, no fewer than 161 American patents were taken out for systems of selective signals for telephone lines. Most of these devices are extremely complicated, almost all of them are inoperative, and none has come into general use."

FIG. 185.
Toll line wall switchboard.

In view of the more recent developments in telephony, this statement must seem increasingly true. As most party line circuits cover no great length of wire, except in the case of some toll systems entering city exchanges from the outlying villages, where it is more economical to string all apparatus on a single circuit, the adoption of elaborate systems of selective signaling seems almost a waste of money and pains, a doubtful way of saving the expense of separate line circuits for each station.

Toll Lines.—The term, "toll line," is used to designate a trunking line between exchanges, which may have a number of private telephones bridged in its length, or where a party line telephone circuit has its terminals in a central exchange switchboard. This designation is said to be derived from the fact that each separate telephoner on the circuit is charged for every occasion on which he uses the line, just as formerly persons driving on country roads paid a designated fee, at the various toll gates they might pass, for the maintenance and repair of the highway. Also, where the term is applied to a trunking line, as stated above, it is proper from the fact that such lines frequently leading to exchanges conducted by other companies involve an additional "toll", or charge, to the subscribers using them.

Toll Boards. — On the other hand, on account of the exposure of such lines to interference from lightning and other weather disturbances ; on account of the fact that they are liable to leakage in proportion to their length and the number of apparatus bridged on the circuit; and because of the difficulty of properly adjusting the circuit arrangements, when required, so as to connect a grounded line to a metallic, it is necessary to use specially constructed switchboards and employ skilled operators. Toll boards differ from others in having fewer drops—generally no more than ten to a position—including devices for switching repeating coils and other necessary devices into circuit, when needed, and in having arrangements for making such selective or code signals as may be in use to call the various telephoners on the line. Several telephone manufacturers have given particular attention to this branch of exchange work, with the result that any company desiring to install toll-line appliances has a wide variety from which to select. One of the most complete and practical of these devices is shown in Fig. 185, which represents a combination switchboard for local exchanges and toll lines, manufactured by the Western Telephone Construction Co., of Chicago.

CHAPTER EIGHTEEN.

PRIVATE TELEPHONE LINES AND INTERCOMMUNICATING SYSTEMS: COMMON RETURN CIRCUITS.

Common Return Intercommunicating Systems.—The general plan of wiring a number of telephone apparatus, so that each one has its own wire, forming a circuit between itself and any other by using a common return wire as the other limb, is adapted primarily to the needs of manufactories, business houses having a number of separate offices or desks, and to small domestic systems. Such small systems require no central office nor any scheme of selective signaling. Moreover, the danger of being "listened in" is reduced to a minimum, if not rendered impossible; and the expense and trouble of caring for the apparatus is similarly reduced by using a common calling battery.

Signal Devices for Common Return Circuits.—In most of the recent schemes for stringing intercommunicating telephone lines every station is, in fact, both "central" and receiving station. That is to say, means are provided in each apparatus for as readily calling any other station as exist in an exchange, and in the point of receiving there is no further trouble necessary than to answer when the signal rings. This end is attained by some form of switch which at any apparatus can connect the common return wire into circuit with the terminal of the desired station at that same apparatus. Such switching is accomplished either by plugging a jack, as in exchange switchboards, or by making the required connections by some form of sliding contact lever key. Both varieties of switch operate by making contact between a leaf spring, such as is used in switchboard jacks, or else between the lever of a pivoted key connecting to the common return wire and any one of a number of anvils or contact points which is the terminal of some station line or other.

Plug and Switch Key Signals.—Figs. 186 and 187 represent two instruments of similar appearance and construction, one furnished with plug board switch, the other with an ordinary keyboard. The latter is manufactured by the Connecticut Tele-

FIGS. 186-187.—Two types of intercommunicating circuit telephones, showing plug and button contact switches.

phone and Electric Co., and the former, by the Western Telephone Construction Co. They are intended for circuits using a common calling battery and individual station talking batteries. The proper wiring connections are indicated beneath the binding posts at the base of the backboard in Fig. 187. In most plug board telephone sets the plug cord is a conductor to connect that particular station into circuit with any other line; in others it is merely a conventional device intended to accomplish the same result by pressing a leaf spring against its anvil, thus connecting two circuit terminals.

Wiring of a Common Battery Circuit.—The proper wiring connections for a three-station common signal battery system is shown in Fig. 189. At the left of the figure is an ordinary standard desk "phone" with the line wires connected to a suitable "terminal block," which also contains the induction coil. The switch keyboard is on a separate piece, as must be the case with desk apparatus of this type. The second station is equipped with a similar desk set, although only the wiring

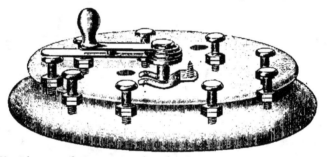

FIG. 188.—Diagram of the construction of a form of button contact switch. The common wire and home apparatus line are wired to the lever, the others, to the binding nuts on the circumference.

connections with the base of the apparatus are shown, and, unlike the one at the first station, the induction coil is enclosed in the base, to which the local talking battery is also connected. The wall set in the third station is one of the kind already depicted. The connections are merely indicated, and the switch may be of either the plug or keyboard type. As may be seen, the connections of each talking circuit have the "common battery wire" as one limb with any one of the three station wires, while the calling circuit includes both the former and the "common return wire." The talking battery of any station may spend along this "common battery wire," because the calling battery itself is then on "open circuit" and emits no current.

Fig. 190 illustrates the wiring connections for an ordinary

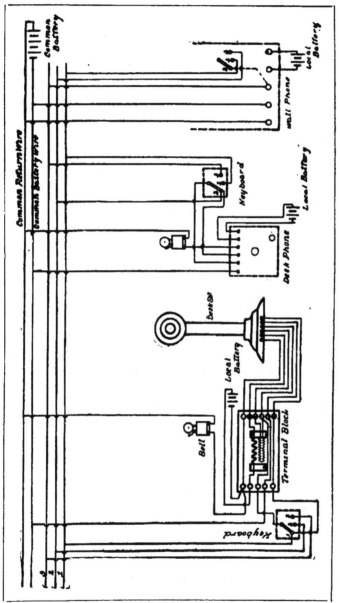

FIG. 189.—Diagram of a three-station, common calling battery intercommunicating system. The two left hand stations are furnished with standard desk apparatus, that on the right with a wall set, as in Fig. 187.

standard desk set with talking battery, magneto-generator and extension or wall bell. Such a system is adapted, first place, for a full metallic line, and when arranged in an intercommunicating system with a common ringing battery the terminals, here shown as connected to the ringer generator, are connected to the station line wire and common return wire, respectively, as indicated in Fig. 189.

Wiring of an Individual Battery Circuit.—As we have seen, all individual line intercommunicating systems have a common return wire which may complete the circuit between any two of them with the wire belonging to the *called station* as the other limb. On a system including three stations, therefore, there would be four wires, as in Fig. 191. This figure shows a series of apparatus, each equipped with its own magneto-generator and transmitter battery. On systems containing a common calling battery, as in Fig. 189, such a three-station line must be equipped with five wires, or two more than the number of stations. If now the telephoner at station 1 on the former diagram wishes to call up station 3, he moves the switch lever of his apparatus to the contact point numbered 3 on his dial, thus bringing wire number 3 into circuit with his apparatus and the common return. If his apparatus is equipped with a magneto the calling circuit is thus made from his station, along wire 3 to line 3, into apparatus 3, where it rings the bell signal, out of apparatus 3 to the common return wire, and thence back to apparatus 1. The talking current follows the same path as soon as the switch hook is raised, as is the case in central exchange apparatus. If, on the other hand, the line be of the common calling battery type, the operation is as follows: Telephoner number 1 moves his switch lever to contact point 3, thus closing a circuit with apparatus 3, and in order to send out a calling signal depresses a button at his apparatus, thus causing a current to move from the common battery along the *common battery wire* until it reaches the wire leading into apparatus 1. Entering there it passes through the apparatus to its connection with wire 3 out upon line 3 to apparatus 3, which it enters, causing

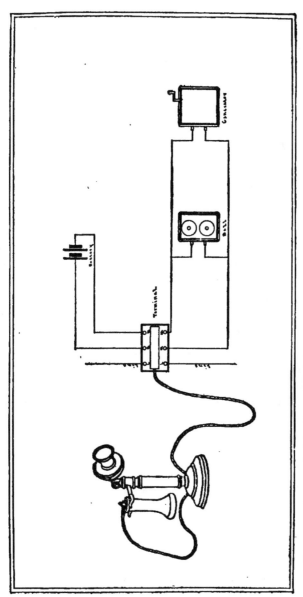

FIG. 190.—Diagram of wiring for a standard desk set and magneto generators; the bells being bridged. This arrangement of circuits is that usually adopted on exchange station instruments.

the bell to ring, and, having passed through the magnet coils, emerges on the *common return wire*, whence it moves to the opposite pole of the battery. The raising of the receiver hook at each station closes the talking circuit, which, as we have seen, uses the *common battery wire* for its return limb on the system of wiring shown in Fig. 189.

Ericsson Intercommunicating System.—Such a five-wire three-party system is shown in greater detail in Fig. 192, which illustrates a system using an Ericsson microtelephone set at stations 1 and 3 and an ordinary wall apparatus, with switch attachment, at station 2. Here the three individual wires are numbered L^1, L^2, L^3 and the *ringing battery* and *common return* wires, $R B$ and $C R$, respectively. The apparatus connections and the wires numbered 1, 2, 3, 4, at stations 1 and 3, correspond to those shown in connection with the diagram of microtelephone circuits in Fig. 130. The apparatus at station 2 is furnished with a switch hook, and is in all respects of the ordinary type, except that it has no magneto-generator. The calling signals from all stations are sent out by merely placing the switch lever on the required contact point and then pressing the push button, P. Thus if station 1 wishes to call up station 3, he places the switch lever on point 3 and depresses the push button, P, of this apparatus, with the result of making the circuit of the calling battery along the wire, $R B$, to the branch wire leading into station 1; through that apparatus to the switch lever, to point 3, to line wire 3, to apparatus 3, where it enters and rings the bell; passing out of apparatus 3 from the other side of the bell magnet to wire 1 of the microtelephone, thence as shown in Fig. 130 of the circuits of such an instrument to wire 2, and thence by the common return wire, $C R$, to the opposite pole of the ringing battery.

When the circuit is arranged to include an individual ringing battery or magneto-generator at each station, it is placed at the point, $R B$ of each apparatus, and the line connection is made with the *common return* wire, $C R$, instead of the *ringer battery* wire, $R B$, which is then omitted, leaving a four-li[ne] three-party circuit, as in Fig. 191. When it is desired to use

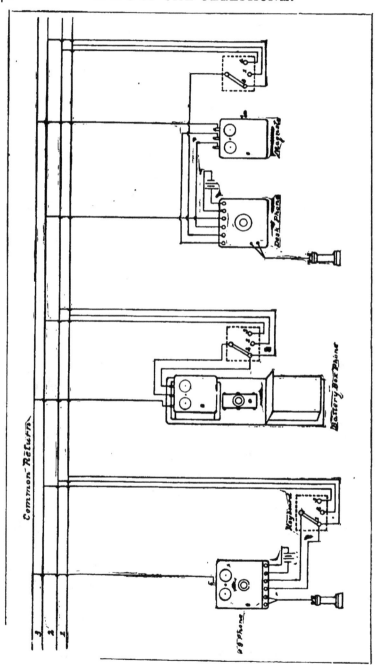

Fig. 191.—Three station intercommunicating system, furnished with magneto call apparatus.

common battery for both talking and signaling, the wires, $T B$, and $R B$, may be connected, thus joining the local transmitter circuits direct to the battery when the switch key, P, is in its normal position.

Device to Insure Ringing of Signals.—According to the wiring of this system, every bell is connected, as shown, with the line wire of that station, so that a calling current coming along its wire, L^1, L^2 or L^3, finds a path directly through the coils of the bell magnets and out to line, the position of the switch lever on any one or other contact point being a matter of indifference so long as the talking circuit is broken and the push button lever, P, is in the raised position, as shown. Thus is avoided the complication found in some of the earlier intercommunicating systems, the necessity of returning the switch lever to the home point after a conversation, a duty that must be frequently neglected through carelessness or preoccupation. Thus the call bell of each station is bridged across the circuit between its proper wire and the common return, so long as the individual talking circuit is not made. The making of that circuit, either by the rising of the switch hook or the act of pressing the lever shown on the shank of the microtelephone instrument, switches out the bell. On the other hand, the releasing of the lever of the hand microtelephone, or the hanging-up of the receiver places the bell once more in circuit, ready to be rung by another call coming from some other station, no matter on what point the switch lever may be left. The advantages of such an arrangement may be readily understood, when it is remembered that one of the most serious problems of early intercommunicating systems was, how to insure the restoring of the lever to the home point after each conversation.

The Ness Automatic Switch Hook. — The difficulties involved in a neglect to return the switch contact lever to the home point after a conversation has been overcome in a different manner by other inventors and manufacturers, whose plan is, in brief, to provide a device for automatically accomplishing this result. One of the most typical and excellent of these is th~

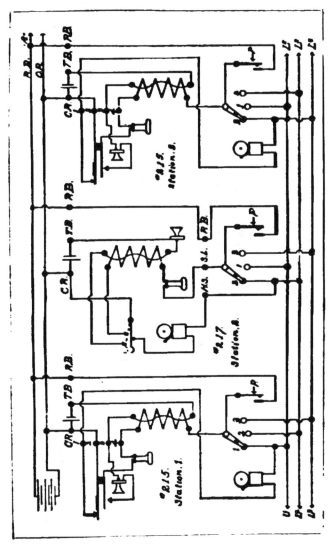

FIG. 192.—Diagram of Ericsson three-station intercommunicating system, two of them equipped with hand microtelephones, the other with an ordinary wall set. Common calling battery.

Ness automatic hook switch manufactured by the Holtzer-Cabot Co., of Boston, Mass. Fig. 195 shows a wall apparatus for a ten-station line equipped with this device, and a magneto generator. In other points there is no departure from the general type of switching telephones, except for the fact that a stop piece for the lever is provided in Ness instruments at the right of the semi-circular row of contact points. This is an essential part of the contrivance, and is intended to hold the lever at the home point when it has automatically sprung back.

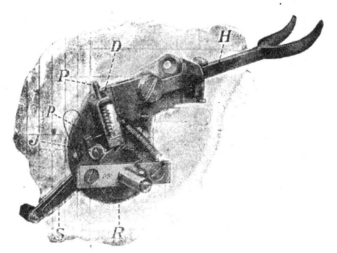

FIG. 193.—Mechanism of the Ness hook switch for intercommunicating telephones.

Mechanism of the Ness Hook.—The mechanism of the Ness switch hook is shown in Fig. 193. Here S is the lever just mentioned, which is normally held in contact with the home point, or right hand contact, on the outside of the door of the box. This right-hand contact connects with line wire 1 in station 1, with line wire 2 in station 2, etc., the order of the points being varied for every station. (See Fig. 194.) If, now, for example, station 1 desires to call up station 8, he moves the switch lever to point 8, as in any other type of switch, being then ready to send out the calling signal. The act of rotating the switch lever turns the ratchet wheel, R, which is held in

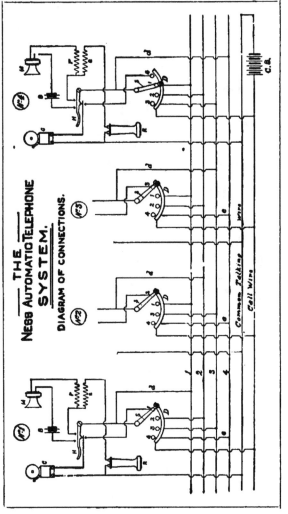

Fig. 194.—Diagram of a four station intercommunicating system operating with the Ness switch hooks.

position at the desired point by means of the pawl, *P*, which is reinforced by a spiral spring, as shown. Having moved the lever, *S*, to point 8, the telephoner at station 1 has his apparatus in position to send out the calling signal current, which he does by pressing the tip of the lever into contact with the semi-circular metal strip shown in Fig. 195 below the row of contact buttons. This act accomplishes the closing of the common battery circuit, as in other systems of intercommuni-

FIG. 195.—Wall apparatus, equipped with Ness switch hook and magneto generator.

cating telephones, sending out the calling current, being connected to the common battery wire, or else makes the proper circuit for the magneto current when no common battery is used.

Operation of the Ness System.—Immediately after sending out the calling signal, the telephoner at station 1 removes his receiver from the hook, which then rises, making the circuit of the talking battery, as in any other type of apparatus. The conversation finished he again hangs his receiver, and the hook is brought back to its lowered position. This act causes the switch handle to fly back to its normal position over the home button by virtue of the mechanism indicated in the figure as follows: On the short arm of the switch hook is pivoted the dog, *D*, which, when the receiver is rehung, engages a notch in the pawl, *P*, and lifts it away from the ratchet wheel, *R*. A spiral spring held around the axle shaft of the ratchet wheel, *R*, then acts, causing the ratchet wheel and the attached handle to be restored to the first position. Meantime the dog, *D*, slips

FIG. 196.—Holtzer-Cabot desk telephone equipped with the Ness switch.

out of the notch on *P*, and allows it to fall into contact with the ratchet in readiness for the next call. In order, however, that the pawl, *P*, may not engage the ratchet before the lever, *S*, has fully reached the normal position, a second dog, *J*, is placed, as shown, being pressed by a spring to a position under the pin, *p*, carried on the pawl, and thus holding the pawl out of engagement until *S* has reached the home point. Then a cam on the under side of the ratchet—it is not shown in the cut—pushes the dog, *J*, away from *P*, and allows the pawl to fall into engagement with the ratchet.

Wiring of the Ness Circuit.—The Ness switch is a very complete and accurate device for achieving the end sought. All parts exposed to wear are made of the best hardened steel, thus ensuring the utmost durability. Best of all, it reduces the number of necessary acts on the telephoner's part to the lowest figure, and leaves nothing dependent upon his memory or carefulness. The full circuits of the Ness automatic system are

Fig. 197.—Globe selective hook switch ; box open showing the slide and spring contacts, also plug in the cover.

shown in diagram in Fig. 194. Here we have a four-station common calling battery line, in which, as may be seen, station 4 is in the act of calling station 1. The circuits, when made, are the same as in the common battery lines just described, being completely indicated at stations 1 and 4, and merely omitted from the drawing at 2 and 3. The wiring diagram also indicates how that the pressure of the switch lever, S, on the semicircular contact piece, D, makes the calling circuit with any station desired from the common battery, *C B*, along the call wire, as shown.

Globe Selective Hook Switch.—The idea of attaching a self-restoring selective switch to the lever of the receiver **hook** is at once brilliant and in the highest degree practical, **since it** takes advantage of a necessary act, that of removing or restoring the receiver, to accomplish several results.   Another **device of** this description is the automatic switch of the Globe Telephone Manufacturing Co.   It may **be** briefly described as follows: The hook lever is of the long shank variety, pivoted at the end opposite the receiver hook. At a point midway on the shank is pivoted a vertical flat metal rod, carrying on each side a number of lugs, each separated from the other by a gap or notch.   Directly behind this vertical rod or slide, although not touching it, is another metal rod, or plate, which forms one terminal of the common calling battery circuit. On either side of the slide

FIG. 198.—Holtzer-Cabot "single point" apparatus for use with Ness switch on the central office set.

is fixed a number of metal leaf springs, as many springs on each side as there are lugs just mentioned, and each of them forms the terminal of some particular station line.   Moreover, in the cover of the box containing the switch hook, vertical slide and rows of springs, are two rows of holes, each one directly over some particular spring.   While the receiver is on the hook the springs are directly opposite the gaps or notches; when it has been removed and the hook shank rises, they are opposite the lugs.   Now to call any desired station the telephoner inserts a plug or pin in the corresponding hole, thus pressing the spring behind it into contact with the battery terminal plate. This makes the calling circuit in the same fashion as in the previously explained systems. He then removes his receiver from the hook, thus allowing the shank to rise and draw the vertical slide upward, so that the lugs are opposite the springs. On releasing the pressure on the spring plug, as just mentioned, the spring makes contact with the back of the lug opposite it, thus making electrical connection between the line it represents and the home apparatus. One advantage claimed for this device is that, when in a factory, for example, the office wishes to call up any number, or all of the apparatus in circuit, it is necessary only to plug several springs instead of one, and thus send out a ringing current to all at once, bringing all into telephone circuit by the act of removing the receiver. On restoring the receiver to the hook the lever is again depressed; the notches again come opposite the springs, releasing any of them that may be in contact. The normal or resting position is thus automatically restored, and the apparatus is always ready to respond to a calling current from any other station.

Switch Circuits.—Any of the forms of switching device described above may be used on the central station apparatus, as in a hotel office, to make the circuit with ordinary wall telephones in the several rooms. Thus if a switch be attached to the office apparatus only, and all the others be such "single point" sets, as is shown in Fig. 198, connection may be made by pressing the button on the wall set, which act operates the office signal.

FIG. 199.—Ericsson's hand microtelephone.

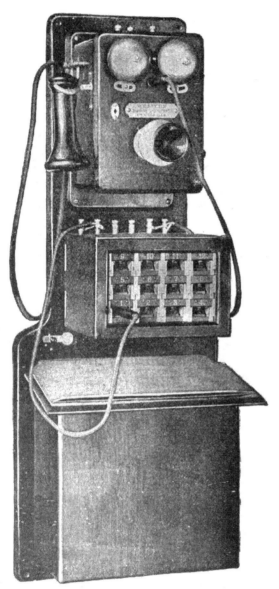

FIG. 200.—Combined station apparatus and switchboard for use in small exchanges or in hotel systems. The drop-jack here used is the one described on pages 156-158.

CHAPTER NINETEEN.

PRIVATE TELEPHONE LINES AND INTERCOMMUNICATING SYSTEMS: FULL METALLIC CIRCUITS.

Full Metallic Circuits.—Many telephonists strongly urge the superiority of full metallic circuits in intercommunicating systems, short and long, both because that by this means perfect privacy may be maintained, and also because, particularly where there is a large number of instruments in the system, more than two of the number may be in telephonic communication at the same time. Fig. 201 gives a diagram of such a full metallic circuit, common battery, intercommunicating system, including four apparatus. The home line wires of each station have their terminals at the left-hand side of each particular apparatus and the battery wires at the right. The terminal points of the other three stations are conveniently arranged midway on the front of each box. The line thus‘ equipped illustrates the system of the Century Telephone Construction Co., one of whose instruments is shown in Fig. 202. As will be noticed, the switch lever makes a double contact at every station point—one with each button of the line to be called, by this means making circuit between both of its terminals and the station line.

The Century Switch.—The lever is constructed with a balanced bearing, is double in form and makes a contact of sufficient weight to keep the points bright, thus insuring perfect electrical conduction. The internal wiring of the apparatus is such that, no matter at what point the lever is left, the station may be readily called by any other. The calling current is transmitted from the common battery, through the calling apparatus, to any desired station by pressing the button shown at the base of the box, as soon as the proper contacts have been

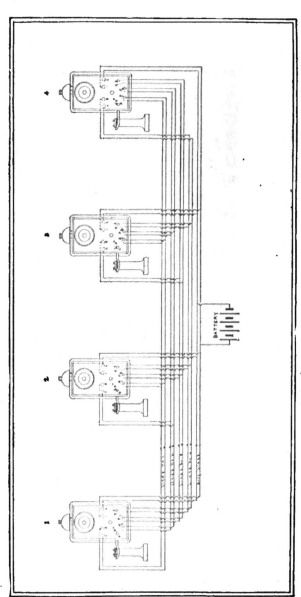

FIG. 201.—Four-station full metallic intercommunicating system with common battery for calling and talking circuits.

made, thus bridging in the lines of the common battery. The talking circuit is made in the usual manner on the removal of the receiver from its hook. Apart from the principle of double connections and double contacts, the switching apparatus is arranged substantially like any other of the sliding pattern

FIG. 202.   FIG. 203.

FIG. 202.—Wall apparatus for use on a full metallic intercommunicating system, as shown in Fig. 201, equipped with double contact switch.
FIG. 203.—Standard desk apparatus for intercommunicating system, showing construction of the Century double contact switch.

adapted for use on single-wire common return systems. Fig. 203, showing a standard desk apparatus, equipped with this form of signaling device, exhibits to good advantage the double construction and double contact of this switch lever, also showing its balanced working on the metal ring within the rows of contact studs.

The Plummer-Monroe Switch.—Another switching device adapted to the requirements of full metallic intercommunicating systems is the Plummer-Monroe switch, which is supplied with some of the apparatus of the Ericsson Telephone Co. It is an ingenious adaptation of the principle of exchange switchboard

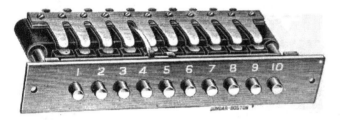

FIG. 204.—Ten-point Plummer-Monroe switch, showing relation of springs and plungers.

spring jacks, with the difference that push buttons are used instead of cord plugs. Fig. 206 shows a telephone apparatus equipped with the Plummer-Monroe switch, in connection with the Ericsson receiver and transmitter already described. As may be seen, ten lines are provided for, but any required number may have their terminals in one station apparatus by placing the push keys and spring jacks in superposed tiers or rows, like the one shown in end view in Fig. 205.

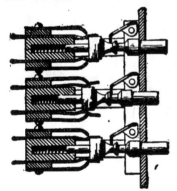

FIG. 205.—Three-tier Plummer-Monroe switch, showing details of construction. The middle plunger is shown pushed in, being held in place by the detent strip hinged on face plate.

Fig. 204 gives a good general idea of the internal construction of this type of switching device. Behind and at either end of the metal face plate may be seen a post connecting it to a cross strip or bridge piece, to which is secured a metal strip carrying ten leaf springs or contact fingers. Below these, and normally insulated from them, is the same number of separate jack springs, each suitably notched or bent, like the springs of switchboard jacks. These are the line terminals.

Each push key carries on its further end a square piece or block as shown, so that when any key is pushed in, the corresponding notched spring is engaged and forced into electrical contact with the contact finger immediately above it. A similar notched spring, in contact with the lower side of each such key block, is engaged in the same fashion, thus making circuit for both sides of that line with the home apparatus represented by the outside, upper and lower strips carrying the contact fingers just mentioned. Between the posts shown to the rear, and running parallel to the face plate, is to be seen a metal piece, which represents a strip or frame pivoted to the blocks to the left of number 1 and to the right of number 10. This is the "detent strip," and its office is to engage a catch just forward of the square contact piece on each plunger, thus effectually holding it in position after it has been pushed into contact with the jack springs. Each plunger is normally held forward by a coiled spring behind it,

FIG. 206.—Station apparatus equipped with a ten-point Plummer switch.

which is within a suitable socket in the bridge piece, and electrically insulated from the jack springs. Thus the act of pushing any button inward causes the detent strip to ride upward on a conical portion of the plunger, just to the rear of the catch just mentioned, thus causing every engaged plunger to be released and be forced outward by its spring. A slight push on any button, moreover, may release any inward pushed plunger without locking in that one, consequently leaving all in the normal position. However, the internal wiring of the apparatus is such that a calling current may enter and ring the bell no matter what plunger is pushed inward. The first act of the telephoner on receiving a call is to push the button corresponding to his station number and then to remove his receiver from the hook, thus making the talking circuit. The advantage of this device is its simplicity of construction and the ease with which it may be operated. .

The Couch & Seeley Switch.—Another switch of somewhat similar construction is shown in Fig. 207. It is manufactured by the Couch & Seeley Co. of Boston, and differs from the Plummer switch in the fact that the plug body of each plunger

FIG. 207.—Couch & Seeley plunger switch.

is conical, instead of square. The detent strip for holding the plug in place after it has been pushed in is swung on the bridge piece at the rear, instead of on the face plate. There are two bus bars for the ringer generator, so that when the button is pressed in it connects the battery to line and transmits the signal. The construction of this switching device permits the removal of the plungers and release levers with the face plate so that the back parts and springs may be readily reached for necessary repairs, or any of the plunger plugs may be removed separately.

Hotel Systems.—Another application of the full metallic line telephone system is to hotels, where, as in some factories, the main office is a sort of exchange, or central station, which may call or be called by any apparatus in the circuit, although generally containing no provision for connecting any two of them in telephonic communication. As the use of telephones in hotels is constantly increasing in popularity, and is recognized as the readiest method for enabling a guest to communicate his wants to the office, a brief notice of such a system is quite in place. Some hotels are equipped with a form of step-

by-step teleseme, consisting of a push button to give the number of the calling room on the office annunciator, and a transmitting dial on which the separate "make" points correspond each to some need of the guest, from an extra blanket to the bill of fare. A similarly marked dial in the office apprises the clerk of the guest's desire. Such a system, however, while having the advantages of all automatic signal devices, is only a little better than the old-fashioned push call for the bell boy, and is immeasurably inferior to a good telephone installation.

FIG. 203.—Common pattern C. & S. jack switch for removable plugs.

Telephone System for Hotels.—Fig. 211 gives a diagram of a hotel telephone system, on full metallic circuits, wherein each separate room circuit terminates in such a combined needle annunciator and single-cord switchboard as is shown in Fig. 212. A separate wire supplies the energy from common calling and talking batteries, the circuit of the former being made by pushing the button shown at the base of the apparatus both in the diagram and in Fig. 209, which gives a view of the kind of telephone set mounted in the separate rooms. The circuit of the common talking battery, which, as a moment's reflection will show, must be separate from the other, as being connected to a different return wire through the annunciator switchboard, is made in the usual way, when the guest and the clerk in the office have both removed their receivers from the hooks. A guest, therefore, desiring to call up the office, pushes the button at the base of the transmitter box of his apparatus, thus closing the circuit of the calling battery and causing the annunciator needle corresponding to his room number to announce his call. The clerk then inserts the plug in the jack similarly numbered, thus establishing connection with the office "phone," and, by removing his receiver from its hook, makes the talking circuit between the two through the talking battery. The process may be reversed, so as to enable the office to call any desired room by simply inserting the plug in the corresponding jack, and pushing a button to send out a current which rings the bell

FIGS. 209-210.—Types of hotel system apparatus : the first for the individual room stations the second to be used in connection with the office annunciator switchboard.

of the room apparatus. The annunciator switchboard shown, and also the diagram of wiring, illustrate the hotel system of the Century Telephone Construction Co., of Cleveland Ohio. The hotel office "phone," shown in Fig. 210, which, as may be understood from the wiring diagram, lacks both the call bell and the calling push button, is the type of apparatus supplied by the Ericsson Telephone Co., of New York, and is furnished with their coal-grain transmitter, previously described, and the polarized ring watch-case receiver first introduced in Europe by Ericsson.

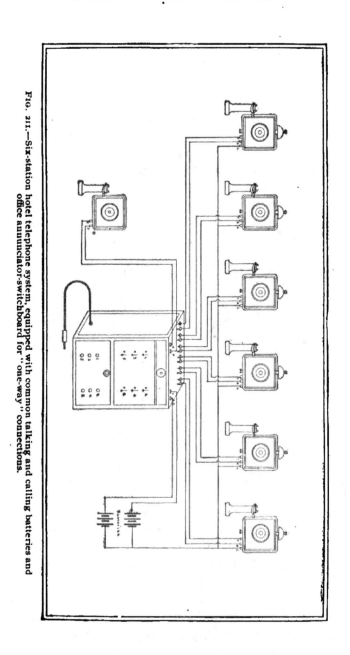

FIG. 211.—Six-station hotel telephone system, equipped with common talking and calling batteries and office annunciator-switchboard for "one-way" connections.

Another type of hotel or factory office switchboard apparatus is that manufactured by the Western Telephone Construction Co., of Chicago, and shown in Fig. 200. This is an ordinary magneto calling apparatus, with the exception that, in place of the usual transmitter arm and coil box, it carries a small drop and jack switchboard. The figure shows an installation of

FIG. 212.—Annunciator-switchboard for hotel systems.   By plugging a jack talking connections may be made between any room apparatus and the office.

twelve drops of the type already described, and calling or answering connections may be made between the apparatus and any other on the line by inserting the plug in the required drop jack. Also, any two stations may be connected as in the ordinary switchboard exchange. A conveniently located switch key enables the clerk or operator to send either a magneto calling current to the other station or to make the talking circuit as is necessary.

CHAPTER TWENTY.

LARGE PRIVATE SYSTEMS AND AUTOMATIC EXCHANGES.

Limited Capacity of Private Telephone Systems.—As a general rule not more than twenty or thirty telephone apparatus may be included in an intercommunicating system, either with common return wire or full metallic circuits. The reason lies in the very obvious fact that even if an indefinite number of stations should be wired to the ordinary apparatus switches, such as have just been described, there would be continual confusion and interference on lines apt to be busy. The alternative, therefore, is to have a private central exchange, with a switchboard arranged to connect the lines of any two stations in the system. Such objections hardly hold good for hotel and large business office telephone systems, which have a "one-way" connection with the main office, since such systems are, in fact, dependent on a central office.

A Private System of Exceptional Size. — A notable exception seems to be presented in such an intercommunicating system as was, some months since, installed by the Stromberg-Carlson Telephone Manufacturing Co., of Chicago, in the offices of the People's Gas Light and Coke Co. of the same city. In order to connect the numerous desks in the book-keeping department and give telephonic communication with the managers' offices and general storerooms, all without the use of a private exchange, it was necessary to provide intercommunicating lines for nearly one hundred apparatus. The problem seems to have been satisfactorily solved by connecting each desk to every other one with which it could possibly need to have communication, and arranging the circuits to be controlled by a system of plug switches. On several of the desks are plug switch boxes with a capacity as high as sixty points; that is to say, providing for switching connections with sixty other apparatus. The wiring is so arranged that the various managers' desk sets may also be put into telephonic circuit with the public exchange by proper plug contacts.

Automatic Telephone Exchanges.—There seems to be a well-defined tendency among telephone users, particularly in the matter of installing private plants, such as has just been described, to avoid the use of an ordinary switchboard exchange, and render the system, as far as possible, automatic. Thus it is that a number of inventors have set themselves to solve the problems involved, with the result that several automatic exchange systems have been put upon the market within recent years. As must be understood, with very small reflection, the most likely method of operating such exchanges is by some variation of the step-by-step principle, which, as we have seen, is the most feasible yet found in the domain of party line selective signals. As a matter of fact, an automatic exchange is only an application of the selective signal idea, so extended as to meet the needs of a larger number of stations than are usually included in an intercommunicating telephone system.

The Strowger Automatic System.—One of the most successful systems of automatic exchange is the Strowger system, which is installed in several American cities. A Strowger exchange in Atlanta, Ga., has as many as 500 subscribers, and estimates an average of twelve calls daily for each one. Another, recently installed in New Bedford, Mass., began business with about the same number, and has been in successful operation for several months. The mechanism of the system is exceedingly complicated, and could not possibly be described within limited space. Briefly, however, we may treat it as follows: The line wires at the central station are arranged in rows in definite order. Over the terminals of branches from these, from the exchange apparatus of each subscriber, is placed transversely a shaft, which, by a suitable arrangement of relays, ratchets and levers, has both a rotary and a longitudinal motion under impulses coming from a common battery. Each such shaft carries a series of arms set upon it spirally; that is to say, occupying different relative radial positions on its circumference. The wires are divided into groups of tens, thus, 101, 111, 121, 131, 141, etc., or 102, 112, 122, 132, 142, etc., and each shaft has an arm corresponding to each such group. Thus a contact between the shaft and the desired line wire may be made through one of these radial arms, according as the switch mechanism at the subscriber's apparatus is operated.

Subscriber's Circuit-Making Apparatus.—Each station apparatus has four keys—one lettered H, for hundreds; a second, T, for tens; a third, U, for units; and a fourth, R, for release. The wiring in connection with these keys is so arranged that circuits are made from the central exchange grounded battery over either limb of the talking line alternately to ground at the subscriber's station. Supposing a given subscriber wishes to call up another, numbered 253, he will press key H twice, thus causing the shaft of his switchboard apparatus to rotate through the space of two make points or teeth, at the same time closing and locking a circuit connection through a second relay over the other limb of the line. He then presses key T five times, thus causing the shaft to be moved longitudinally, through five steps and locking circuit connection, through a third relay. When key U has been pressed three times one arm of the shaft is brought into contact with wire 253, and communication may be had by ringing his magneto bell to call up subscriber 253. If his apparatus bell does not ring he knows that 253 is engaged, and accordingly awaits another opportunity. When the conversation is completed he presses the key R, thus operating a restoring mechanism at "central," and leaving his switchboard apparatus in position for another call. The later apparatus of this system have a simple dial arrangement, instead of buttons, for transmitting the current impulses, and the method is to send out three distinct sets of impulses for every call: the first for hundreds, by turning the dial to the

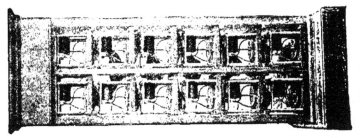

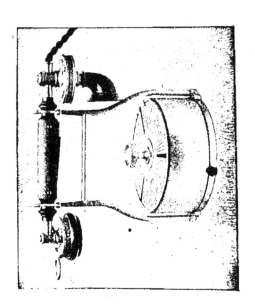

Figs. 213, 214, 215.—The Clark automatic switchboard, transmitting dial, subscriber's receiving relay for switching at the exchange and frame for mounting subscribers' relays.

required number from o to 9; the second for tens, by turning the dial to the required number from o to 9; the third for units in the same fashion. Each separate turn energizes the proper relay the required number of times, then locks connection for the next set of impulses.

The Clark Automatic Exchange.—A much simpler system, adapted for use on lines of not more than 150 stations, is the one manufactured by the Clark Automatic Telephone Switch-board Co., of Providence, R. I. This may be described as a thoroughly representative step-by-step mechanism, and has the advantage of combining simplicity of construction with exactitude of action. Each subscriber's station has a call-transmitting device, consisting of a dial around whose circumference is a succession of numbers, from 1 to 75, corresponding to the numbers of the stations in the system. This is shown in Fig. 213. When one subscriber desires communication with any other he pushes in the button shown at the side of the trans-mitting dial, thus releasing a locking arrangement which normally holds the indicator mechanism at any desired point. He then turns the dial by the knob at the center until the desired number is opposite an indicated point. By this operation he causes a circuit-making arm to pass over as many "make" points as there are between the start and the required number, thus alternately energizing and de-energizing an electro-magnet at the central station a corresponding number of times. This electro-magnet is shown in Fig. 215, which gives a good general idea of its operation. There is one such automatic switch, or "receiver," to every subscriber's apparatus. To the armature of the magnet is attached an escapement, as shown, which is normally held away from the pole by a small coiled spring. Each time the armature is attracted the escapement moves the ratchet wheel forward one-half a tooth, each release making the other half under the force of the spring. The contact buttons arranged around the circumference of the wheel are the terminals of the various apparatus wires in the system, and each movement of the ratchet through the space of one tooth makes contact with one of these circuits. The forward movement continues until the required number is reached, for since the number of make points on the transmitting dial and the contact points on the ratchet wheel correspond exactly, the movements will always be in unison. The magnet takes energy from a battery included in the circuit. When the conversation is completed the subscriber again pushes in the plunger on his transmitting dial, and moves the numbered dial through the remainder of its revolution until it is restored to the neutral point from which it started, and is ready to start again. Fig. 214 shows the form of frame for mounting the subscriber's switching apparatus at the exchange.

CHAPTER TWENTY-ONE

DEVICES FOR PROTECTING TELEPHONE APPARATUS.
FROM ELECTRICAL DISTURBANCES.

Protective Devices.—Beginning the study of line wiring
at a terminal station, the first things to attract the attention are
the devices constructed to afford the apparatus protection from
the damaging effects of lightning and sneak currents. The need
of the former may be readily understood even by those who
have little or no electrical knowledge; since, as all know, steel
instruments and electrical contrivances, when in a position
exposed to lightning, are apt to attract the bolts and suffer
damage accordingly. The latter danger, exposure to sneak
currents, due to the lines crossing with power wires or other
conductors of electricity, is quite as grave. Such currents,
entering a station apparatus or exchange, are liable to "burn
out" all the magnet coils and other appliances by so heating
the wires that their insulating covers are charred and rendered
useless. To guard against such disaster devices are installed
which are able to break the circuit when charged by a current
of more than a specified amperage.

Fuse Wires.—In order to thus break the circuit under the
stress of over-heating, wires of fusible metal—usually an alloy
of tin and bismuth in such proportion as to melt at a desired
temperature—are interposed at proper points in the circuit, as,
for example, at the entrance of a station apparatus. The
arrangement of such a fuse is shown in Fig. 216, where the line
wire and the direction of the current are indicated by the
arrows, the fuse wire being between the two connecting points.
If a current of more than the safe strength come along the line,
the fuse is melted, or "blown out," as its resistance to the
current generates heat, and the circuit is thus broken. Owing

to the constant necessity of replacing blown-out fuses, such wires are usually attached between plates, or clips, of harder metal, such as are shown in Figs. 217 and 218, and these are inserted in circuit, either by screws, as in Fig. 222, or between leaf springs.   Each such fuse plate or clip has a flange, intended to hold a sheet of mica of proper length as mounting to the fuse wire, as shown in the same figure.

FIG. 216.—Fuse wire attached to connectors.

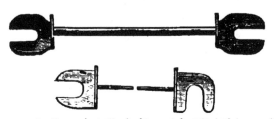

FIGS. 217–218.—Fuse wires attached to metal contact plates or clips.

Heat Coils and Circuit Breakers.—While it is generally a sufficient protection against sneak currents to arrange the fuses as shown in the foregoing figures, the requirements of line work, in pole and station terminals, frequently demand some additional device which will insure the breaking of the circuit the moment the heat is of a sufficient degree to melt the fuse wire.   Such devices are desirable from the fact that a badly arranged fuse—one not of the proper length or diameter— under certain conditions, such as will enable a short fuse to heat its metal clips by radiation, and thus increase its current-carrying capacity, will frequently admit of damage before the circuit is opened.   It is also very difficult often to arrange a fuse to blow off with the desired degrees of current.   To meet these conditions some inventors have enclosed in a coil of German silver a short hooked pin, the two being soldered with a few drops of fuse metal.   A strong spring is then attached to he hooked pin, and the moment the fuse metal is heated to the melting point by a current in the highly resistant coil, it is pulled forcibly from its socket, thus cutting the circuit. A device of this description (Ericsson's heat coil) is shown in Fig. 219.

Tubular Fuses.—Under the general head of tubular fuses, we may place such devices as provide for surrounding the fuse wire with a sheath or jacket of non-conducting material. A representative protection of this kind is the Cook tubular fuse, a group of which is shown in Fig. 220, arranged to be fitted to a cable terminal. The case is a cylinder of hard rubber, carrying a threaded bore in its length. Into one end of this is screwed one terminal brass cap. Into the other end is screwed a brass plug with flanging end to hold in position a coil of German silver wire, which coil connects the two brass terminal caps, passing through a hole running in the length of the tube parallel to the screw-threaded bore. Through the center of the screw-plug is a longitudinal hole, into which is secured, with a low fusion solder, a brass rod attached to the terminal brass cap, which closes that end. A current of unusually high amperage passing through the coil of German silver wire produces sufficient heat to melt the fusile solder, and allows the cap and rod to be drawn from the tube by tension springs attached to either end, thus breaking the circuit through the heat coil.

Fig. 219.

Lightning Arresters.—A simple form of lightning arrester, still used on some telephones, consists of three serrated metal plates, which are arranged in a staggered row on the top of the generator box, each being secured in place by one of the binding posts of the apparatus. The third, or middle, binding post is connected to ground ; the other two are the terminals of the line wires. As these plates are not in contact—contact would interfere with the operation of the apparatus—the method of securing efficient protection from lightning is to place a metal plug between the two line terminals, thus short-circuiting the apparatus ; or, to connect either or both with the grounded plate by similar plugs.   The theory of this simple device is that the lightning will not affect the apparatus when it is short-circuited, rather preferring to jump the short distance to the grounded plate and pass off harmless.   This result, however, does not always follow, as lightning is one of those things that often

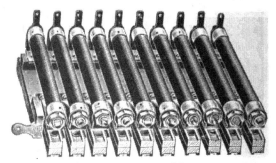

FIG. 220.—Strip of Cook heat coils arranged for use on a cable terminal.

seems to be an exception to all rules.   Another difficulty is found in the fact that should a subscriber neglect to remove the plug at the end of a thunder storm his instrument is rendered inoperative.   The line circuit being completed between the plates at the top of his apparatus, no current will enter.   Such neglect has frequently worked great inconvenience, disabling an entire line when a number of instruments are included in a bridging circuit.

Carbon Lightning Arresters.—A much more efficient method of grounding a lightning charge short of the apparatus to be protected is by the use of carbon arresters.   As produced by nearly all telephone manufacturers, the carbon arrester consists of two flat blocks of carbon, between which is placed a thin sheet of mica.   Such pairs of carbon blocks are attached to each terminal of the line, one on each being connected to line, the other to ground.   The direct line to the telephone apparatus is, however, not broken on either limb, the carbons forming a branch circuit. The theory is that the lightning current will pass through the small hole in the mica strip between the carbon blocks, and follow the line of least resistance to ground. In practice this theory seems amply warranted. The Sterling Electric Co. have introduced a further protective feature:

Fig. 221.—Sterling combined heat coils and carbon arresters for protecting a subscriber's station apparatus; also long tubular fuse for protection on the line outside the subscriber's station.

not only perforating the mica plate at a point midway on its length, but also inserting in a hole in one of the blocks a small drop of fusible metal. Under a high heat pressure this metal will melt, thus assuring perfect electrical connection between the two blocks, and grounding the line. The fuse and carbon protectors are frequently combined in one instrument, as shown in the accompanying figures, thus assuring complete protection to the apparatus from all electrical disturbances.

Line Protectives.—In well-constructed telephone lines every exposed point is protected as thoroughly as possible from lightning and sneak currents.   Thus in addition to the combined fuse and carbon arresters attached to the telephone apparatus, outside protectives are provided to doubly assure the result.   In Fig. 221 are shown the station protectives used by the Sterling Electric Co.   Within is the tubular fuse and carbon, and at the joining of the line outside is a long tubular fuse.   A current that will pass through the one must meet the other. Similar tubular fuses are provided at all cable terminals, as shown in Fig. 220.

FIG. 222.—Fuse and carbon protector for subscriber's station apparatus.

CHAPTER TWENTY-TWO.

THE GENERAL CONDITIONS OF TELEPHONE LINE CONSTRUCTION.

Construction of Lines.—The subject of telephone lines includes a large number of details—such as proper constructions, testing, repairing—which make it a profession by itself, to be mastered only by practical experience and careful training. In addition we may include under this general head the matters connected with terminal devices and the various protectives used on lines and apparatus and at central exchanges. In entering upon a review of the matter we must bear in mind that there are two distinct kinds of lines in telephone practice, as also in telegraphy; lines with a ground return and full metallic lines, in which both limbs of the circuit are wires of the same material and dimensions. There are also two ways of constructing lines connecting terminal stations; stringing them on poles in the usual familiar fashion, and enclosing them in cables, which are run through properly constructed underground conduits. Very frequently also wires hung on poles are bunched into cables of the same description and suspended by hangers. The majority of lines also are strung on poles through one portion of their length and buried in conduits through another portion. All these points of construction involve the use of special devices, which will be explained in place.

The Conditions of Line Construction.—By such devices as have been described in the last chapter the apparatus at stations and exchanges may be protected from danger by foreign currents. There are, however, so many precautions necessary to the end of securing a thoroughly quiet and serviceable telephone circuit that we may venture to assert that the science of telephone line construction is very largely summed up in the knowledge of how best to overcome the electrical difficulties and obstacles that must be met. In the first place, the question of the material and dimensions of the line wire must be carefully considered, both from the standpoint of durability and from that of conductance. For, when we have overcome the difficulties incident to the specific resistance of a metal by enlarging the diameter of the wire, we are met by a new one quite as serious. By enlarging the diameter we have increased the total surface, or circumference, area of the line, hence increasing its electrostatic capacity for electrical condensation with the earth as the other charging surface and the intervening air as the insulator. This condition, which is a wonderful deterrent to successful telephonic transmission, may be largely neutralized by using taller poles; but here also is a difficulty, as above a certain height poles are exceedingly liable to be uprooted in a storm, with the obvious result of disabling the entire system. In order to avoid another cause of electrical waste it is essential that all joints be tight and secure, soldered where not otherwise protected, in order that the ends of the two wires so connected may not become a kind of loose contact microphone to the disadvantage of good conduction of speech. Finally the disturbances due to induction, both electro-magnetic and electrostatic, from other wires, telephonic, telegraphic, or power, when these cross or are strung near the line, have to be neutralized. It is thus easy to see that the matter of line wiring for a telephone circuit involves many other considerations than merely attaching a wire of any metal to a series of poles of sufficient strength. Every phase of the question must be carefully considered and planned before work is begun on any given line.

General Inductive Disturbances.—One matter which should be understood by every practical telephonist relates to the phenomena characteristic of the constant shifting and reversals in the magnetic properties of a conductor carrying the telephonic current. As we have seen in a previous chapter, ·y current-bearing conductor, whether insulated or not, is unded by a whirl of magnetic force, the lines of which run at right angles to its length.   As we might naturally conclude, the directon in which this whirl of force moves is in accord with the direction of the electrical current on the wire.   Hence it is that every time the current is interrupted, diminished or alternated, there is a corresponding alteration in the magnetic conditions of the wire, which demands readjustment or a new adjustment of the magnetic field.   This process, of course, demands time and a certain expenditure of electrical energy, and for this reason constitutes a species of false or apparent resistance, comparable to the resisting property of water in a pipe when met by any sudden change in its rate of motion, either as an effort to start or to check its flow.   The tendency of matter to maintain any condition is what physicists term "inertia."   On the same figure, then, do we speak of magnetic impedance or self-induction of an electric circuit as electrical inertia.   It is most powerfully illustrated in a coiled conductor, in which, as we can readily understand, the lines of each coil cross those of the ones next following, demanding possibly numerous adjustments of their relative positions and directions, which explain the delay in closing a live circuit through such a coiled conductor.   The "magnetic lag," as it is termed, is vastly increased by introducing an iron core into the coil—making an electro-magnet of it—since the magnetization of the core constitutes a true impedance, the process of demagnetization, on each cessation of the current, gives rise to the condition known as "retardation."   This is the principle applied in the long-wound magnets of bridging bells, which form an efficient bar to the rapidly alternating telephonic currents.   Indeed, alternating currents in general are liable to suffer from the resisting action of self-induction in direct proportion to their frequency.

Inductive Disturbances in Telephone Lines. — In telephony the effects of cross-induction between different lines and of self-induction are found in the "absorption" of the higher overtones of the voice, which renders the sound received both small in volume and indistinct.   The speech-bearing current is extremely complex, and, as transmitted by the modern tele-

phonic apparatus, alternates with high frequency and an indescribable multitude of wave lengths and shapes due to the
blending of the overtones of the sound waves with the fundamental notes of characteristic volume and timbre. The transformation of the sound impulses into electrical impulses gener-
rates a correspondingly large variety in phase—that is to say,
produces a series of waves of varying length, speed and frequency, which must inevitably involve the loss of much of the
original vocal quality—many waves being "choked off" before
completing their phase—although perfectly carrying the fundamentals and many of the lower overtones. Thus it is that conditions which permit of the ready transmission of regularly
alternating currents intended for power use are important interferents in telephony, which deals with alternating currents of
indescribable irregularity.

The Electrostatic Conditions of a Line.—Another form
of interference in telephone lines is due to derived induction
from other current-bearing circuits, which causes "cross talk"
and such foreign disturbing noises as arise from proximity to an
alternating current power wire or a telegraph line. As has been
several times suggested, every line is in fact a condenser, with
the wire for one charged surface and the earth, or some other
near conductor, as the other. This involves a continued necessity of recharging and readjusting the polarities every time the
current alternates. In the section on electrical condensers we
stated that the basic fact of condensation was a produced difference in potential between the electrically charged surfaces.
Thus it is that to change the polarity of the source or to alternate the current on the line involves a shifting in the potential,
which means not only a change in this respect in the conducting
surface immediately charged, but also in every other such surface in the system affected by induction from the line.

Electrostatic Induction.—It has been demonstrated that
the disturbing influences of "cross talk" and other noises are
due principally not to electro-magnetic induction but to electro-

static induction, such as is seen in the process of "charging " a condenser. The series of experiments by which this fact was established by Mr. J. J. Carty, inventor of the "bridging bell," consisted in running two grounded circuits of equal length at equal distances apart, arranging an ordinary transmitter apparatus at the line extremity of the first, and three receivers, one at either end and one in the middle, in the second. Words spoken into the transmitter on the first circuit could be distinctly heard in the two end receivers of the second circuit, but not at all in the middle one, thus proving that the inducing current moved either to or from the middle and neutral point, which is an ascertained characteristic of electrostatic charges, and not at all from either end through the whole length of the circuit, as is the case in such electro-magnetic induction devices as the ordinary induction coil. The problem of overcoming the condition is then a simple one, involving merely an observance of the laws governing induction of this variety.

Electrostatic Capacity.—As has been well said in regard to telegraph lines : " When a key is depressed, closing a long telegraph current and sending a signal into a line, it is at least very probable that a portion of the electricity travels to the end of the wire with the velocity of light. But as the wire must be charged, enough current to move the relay may not reach the end for some seconds." The amount of electrical energy required to thus charge, or shift the potential of a line, varies with several other facts, such as the diameter of the wires, the height of the poles, the proximity of other lines, sometimes as the strength of the current, and as the line is grounded or on full metallic circuit. According to the observance of the conditions thus enumerated in the construction of a line, it differs from other lines, exactly the same in other particulars, but under different conditions, in a quality known as "capacity." In one sense capacity is to be determined by the area of the conducting surfaces to be charged; but in the practical aspects of the situation, so far as regards condensers of all descriptions, the rule followed is that it varies according to the thickness of the "'dielectric," or insulating layer, between the conducting surfaces. Thus the same telephone or telegraph line may have different capacities on different sections in its length, any two such sections differing in most of the particulars above enumerated. The rule seems, in this way, to be established that electrical condensation demands a maintenance of uniformity in the conditions that render it operative. The converse follows, therefore, that to disturb this same uniformity, in such fashion as not to interfere with the conducting properties of the circuit, will accomplish the end of neutralizing the effects of electrostatic disturbance. Such shifting must, however, be regular and proportional, in order to balance the line and to avoid substituting a large number of short condensing areas for a few longer ones. The means usually adopted is the arrangement known as "transposition," by which the wires of several circuits are so shifted at proportioned intervals that their relative positions are quite disturbed, and all electrostatic influences are reduced to the lowest point. This system will be fully explained in the proper place.

FIG. 223.—Pocket battery gauge. Such an instrument is indispensable to every electrician but the readings of its records must first be determined by some known standard, since they are purely arbitrary.

CHAPTER TWENTY-THREE.

TELEPHONE POLE LINES.

Pole Line Construction.—While it would be impossible to enter into all the details and give all the figures necessary to a complete understanding of the construction of pole-suspended telephone lines, the general principles may be expressed in a few pages. The first considerations are the strength and durability of the poles, which involve calculations on the size, the cross diameter of both top and bottom in proportion to the height, and of the kind of wood to be used. Again, and by no means of subordinate importance, is the consideration of what proportion of the total length of a pole must be planted in the ground in order to afford a secure hold.

Wood for Poles.—It will be readily understood that the durability of a pole depends largely upon the kind of wood of which it is composed. Moreover, no commercial system of treating the wood can insure its preservation for so long a time as to materially alter the proportional figures on this point. When pole lines are strung in cities, where protection is afforded from many inclemencies of the weather, the material used is generally Norway pine. This is the case because trees of this variety frequently attain the required height. In cross-country construction, where, other things being equal, strength and durability of construction are considerations of equal importance with height, poles are most often of chestnut, cypress or cedar. As estimated by several authorities, the average life of the varieties of wood most commonly used in pole-line construction is as follows : Norway pine, 6 years; cypress, 10 years; cedar, 12 years; chestnut, 15 years.

Dimensions of Poles for Telephone Lines.—The following data, given in Kempster Miller's "American Telephone

Practice," show the dimensions of poles suitable to the requirements of various lines. In general the poles, of whatever
height, must be at least seven inches in diameter at the top.
The diameter at the bottom, and the depth at which the pole is
planted in the ground varies directly with the height.

The figures are as follows :

LENGTH OF POLE.	DIAMETER, 6 FT. FROM BUTT.	LENGTH PLANTED IN EARTH.
25 feet.	9 inches.	5½ feet.
30 feet.	10 inches.	6 feet.
35 feet.	11 inches.	6 feet.
40 feet.	12 inches.	6 feet.
45 feet.	13 inches.	6½ feet.
50 feet.	14 inches.	6½ feet.
55 feet.	16 inches.	6½ feet.
60 feet.	17 inches.	7 feet.
65 feet.	18 inches.	7 feet.
70 feet.	20 inches.	7½ feet.

Preparing the Poles.—To prepare a tree-trunk for use as
a telegraph or telephone pole it is necessary to peel away the
bark as soon as it is felled, carefully shaving down the knots,
and leave it to dry, in order that the sap may be evaporated,
and one cause of decay thus effectually removed. A pole thus
treated will last its full period, particularly if it be painted.
There are several methods of preparing the wood chemically so
as to prevent decay, but the cheapest of them involves an initial
expenditure and complicated appliances hardly warranted for
small poles, and almost prohibitory for the longer ones. The
most familiar of such methods are those known as "creosoting"
and "vulcanizing," both extensively applied in Europe. In
America the usual practice is to give the pole a heavy coating
of pitch over the first six feet from the butt, which serves to
protect it from moisture of the ground.

Cross Arms.—The familiar cross arms for stringing the
wires are usually attached to the poles before they are erected.
They are commonly made from yellow pine wood, generally 3¼

by 4¼ inches square, and are freely coated with good mineral paint as a preservative. Attachment is made to the pole by cutting a *gain* one inch deep and of sufficient breadth to allow the longest side of the cross arm to fit accurately. It is then secured in place by a lag screw, a long wood screw with a square nut head, so that it may be driven into place with a wrench. The cross-arm is further secured to the pole with braces.

Before the cross-arm is set in place the gain is carefully painted with white lead. As it is important that the cross-arms

FIGS. 224, 225, 226.—Insulator pin for attaching to cross arm ; bracket for attaching to pole or other support ; petticoat insulator, to be made of glass or porcelain.

on a line of poles, particularly when there are several on each one, should be at equal distances from the ground as well as being uniformly spaced, it is necessary that some measuring instrument should be used to secure this end. Such an instrument is the ordinary *template*, which is a length of board carrying a pointed block at one end, to correspond exactly with the top of the pole, and also cross cleats nailed at precisely the same intervals below it as it is proposed attaching the cross-arms. The template, laid upon a pole, shows where to cut the gains.

Attachments for the Wires.—The cross arms are bored with holes for the insertion of the insulator pins, which are made of locust wood and threaded at the upper end to attach the glass insulator. According to the number of the pins to be inserted in a cross arm it is made shorter or longer. An arm for two pins is made three feet long, according to the standard usually followed, with holes for the pins at points three inches from either end and a space of 28 inches between them in the center.   The pins are inserted in the cross arm in even numbers, 2, 4, 6, 8, etc., owing to the fact that it is customary to string the two limbs of every circuit, its line and return wires, to the same cross arm on every pole.   According to the standard sys- tem of measuring, a 10-pin arm is made 10 feet long, giving a space of four inches from the last pin to either end, 16 inches between the two center pins, and 10 inches between all others. The standard system is that followed in telegraph-line work, and frequently also in telephone lines.   Of late, however, a tele-

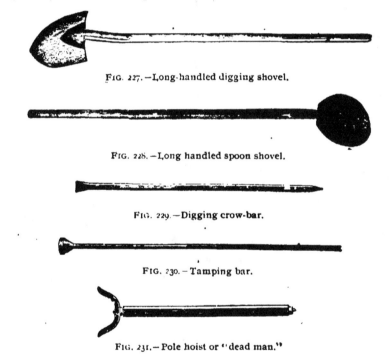

FIG. 227.—Long-handled digging shovel.

FIG. 228.—Long handled spoon shovel.

FIG. 229.—Digging crow-bar.

FIG. 230.—Tamping bar.

FIG. 231.—Pole hoist or "dead man."

phone arm spacing and dimension system has been widely adopted, which differs in some important particulars.   In this system the cross arms are made 2¾ by 3¾ inches square, the end pins are placed three inches from the ends, and on all ex- cept the 2-pin arms they are spaced at 10 inches apart.   Accord- ing to the table given by Kempster Miller, a 2-pin arm may be

24, 30 or 36 inches long, allowing a spacing in the center of 18, 24 or 30 inches, respectively. For four pins the length is 42 inches; for six pins 62 inches; for eight pins 82 inches; for ten pins 102 inches; for twelve pins 120 inches. As may be seen, therefore, the 10-foot arm takes twelve pins instead of only ten, as in the standard measurement.

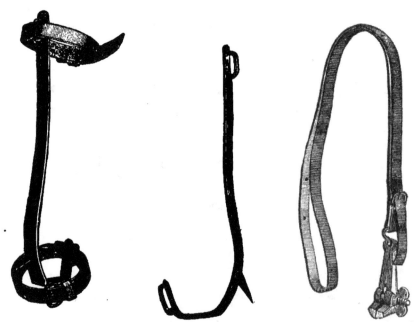

FIGS. 232-233.—Eastern pole climbers, with and without strap for attaching to legs.

FIG. 234.—Portable vise with strap for pulling up the slack in splicing.

Spacing the Poles.—Poles for a telephone line may be placed at intervals varying from twenty to fifty to the mile—that is to say, approximately, from 100 to 260 feet apart. In general the spacing of the poles, like their dimensions, is regulated by the weight of the lines they are designed to carry—the heavier the lines the nearer the poles—and also by their liability to injury from storms and wind in any given locality.

Planting the Poles.—Since each pole on a properly constructed line is sawed to the right length and carefully shaped

before it is finally inserted in the ground, it is necessary that
the holes be dug to as nearly the required depth as possible.
Holes for poles are dug very little wider than their diameter at
the butt, and the depth is usually computed according to the
nature of the soil and the weight of the proposed line, the figures given above being, however, fairly representative of the

FIGS. 235-236.—Two forms of "come-along." The wire is inserted between jaws and is
held fast when tension is applied to the ring.

general practice in this particular.   Excavation, while sometimes accomplished with patent post-hole augers, or even dyna-
mite, is usually done with long-handled digging shovels and
the earth removed with spoon shovels, such as are shown in the
accompanying figures.   Wherever necessary a foundation of
loose stones is formed in the bottom of the hole, and in marshy
or springy ground a basis of concrete and cement is laid, with
filling of the same material around the pole, when raised.   The
poles are rolled to the holes, or carried on hooks similar to those
used for carrying blocks of ice, except for a long handle for
lifting the load at either side. A piece of timber is then inserted
'he hole as a slide to prevent crumbling of the earth as the

pole is slid into place. The end is raised by hand sufficiently to allow the "dead man," or pole hoist shown in Fig. 231 to be placed beneath, and this is moved along regularly as the pole is lifted with pike poles, until it slides into place through the force of gravity. This accomplished, supports are arranged

FIG. 237.—Portable pay-out reel for line wiring.

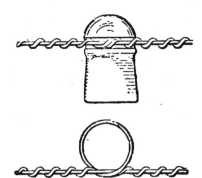

FIG. 238.—Pony insulator and the latest approved method of tying on the wire.

around to hold it in position while the earth is carefully shoveled into the holes and thoroughly packed down with a tamper. In order to secure the pole from over-strain, which might throw it down and break the wires, guy cables are largely employed; these being attached to the top of a pole and secured either to the base of the next pole, to a suitable guy stub or post or to a guy anchor, which is buried about eight feet in the earth and held down by stones and concrete. Guying is most frequently resorted to when the line turns a corner. Then it is necessary to thoroughly secure the poles so that no strain may come on the cornerwise span.

Wire for Telephone Lines.—In telegraph lines galvanized
·iron wire is usually employed, the thin coating of zinc serving
to protect it from the corrosive action of the elements.   Iron
wire is also used to a limited extent in telephone work, but, from
its greater conductivity, lesser capacity, owing to the smaller
exposed surface, its greater durability and lesser weight per mile

FIG. 239.—American wire joint.

FIG. 240.—Block and falls used, as shown, to hold wires for splicing.

FIG. 241.—The McIntire sleeve joint, before and after twisting.

copper wire is far superior.   Iron wire, even when thoroughly
well galvanized—it is coated with a thin layer of metallic zinc
by passing through a bath of the molten metal—is short-lived,
lasting at best only from four to six years, and losing much of
its conductivity long before.   The copper wire used in telephone
lines should be the hard drawn rather than the annealed, and
have a breaking weight of between two and a half and three
times its weight per mile.   Thus the best copper wire of size
oooo (B. & S.), such as is used as the power wire for trolley car
lines, weighs 3,382 pounds to the mile, and will stand a lateral
strain up to about 9,971 pounds.   Wire of size 12, the gauge
most often used in telephone lines, weighs 166 pounds to the

mile and will stand a lateral strain up to 307 pounds, which is its breaking weight. The former has an approximate resistance of one-quarter ohm per mile, the latter of about 5.2 ohms. The following table will give a good idea of the relative merits of the best quality of iron and copper wires of size 14 (B. & S.), which is the one frequently used on short lines:

METAL.	WEIGHT PER MILE.	BREAKING STRENGTH.	RESISTANCE PER MILE.
Iron.	96 pounds.	541 pounds.	49.8 ohms.
Copper.	83 pounds.	193 pounds.	8.7 ohms.

As the data regarding the gauging of wires will be given later we will turn to a consideration of the methods employed in stringing a telephone line.

Stringing a Line.—The erection and guying of the poles of a line as well as the attachment of the cross arms and the screwing-on of the insulator caps are completed before the stringing of the line is begun. It is particularly essential that the pull on poles of a given line be accurately calculated, and that each one be guyed accordingly before the line is strung, in order to avoid the danger of an undue strain upon the wires in attempting to rectify the condition afterward. It is a good working rule that the wires should be subjected to no stress other than the weights of their own spans after they have been attached to the poles.

In stringing the lines either one or the full number of wires may be put up at the same time. When one line only is to be strung the operation consists simply in reeling the wire and running it off from a hand reel, such as is shown in Fig. 237. At each pole the wire is drawn up to its place, pulled out to the desired tension, and attached to the insulator. In the operation of stringing a number of lines at once the method is different. The reels are placed at the beginning of a section, each wire being inserted and secured through a separate hole in a board, which is perforated to correspond exactly with the spacing of the insulators on the cross arms. A rope is then attached to this running board, which is drawn by a team of horses through the stretch to be wired, being lifted over each pole top in turn.

FIG. 242.—Insulated handle side-cutting pliers.

FIG. 243.—Wire-splicing clamps or connectors.

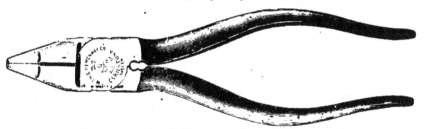

FIG. 244.—Combined side cutting pliers and wire connectors.

When a certain length has thus been drawn out the wires are drawn to the required tension between each pair of poles and secured to the insulators.

Tension and Sag.—In applying tension to the wires as they are strung on the poles it is the rule to allow some sag. To draw them perfectly tight would mean to permit an undue strain when the metal contracts in cold weather. The amount of sag to be allowed varies with different line hangers. A typical case quoted by one or two authorities gives a sag of four inches at the center of each 130-foot span. A more general rule is to make the tension on a wire as it is drawn up between each pair of poles equal to one-third of its breaking weight Thus No. 10 (B. & S.) would be drawn to about 163 pounds and No. 12 to about 102 pounds. The temperature at the time of stringing and the distance between the poles are, however, important considerations in applying tension and allowing for sag.

In drawing out the wire it is customary to use a wire clamp, or "come-along," two forms of which are shown in Figs. 235 and 236. This tool is attached to a block and tackle, or drawn in by hand, and, as soon as the proper force has been applied, the wire is held, while the lineman secures it to the insulator. Another contrivance for this purpose is the pole ratchet, by which the wire is drawn tight and held until attached to the pole.

Attaching the Wires.—Practice has developed a number of methods for tying the wire to the insulators and also for joining the ends of the separate lengths. Many of the earlier methods of tying accomplished the end of obtaining a secure tie by using a wire of smaller diameter than the line wire to pass around the insulator and fasten it firmly. Fig. 238 shows the most approved method of tying in present-day line working. The line wire is first laid in the groove of the insulator, after which a short piece of the same size of wire is passed entirely around to hold it in place, then it is twisted to the line at either side with pliers.

Splicing the Wires.—Appropriate means for joining the separate lengths of wire are of even greater importance. There are several such also. Fig. 239 shows what is generally called the "American wire joint." The method of making it is, briefly, to grip the two wires with a hand vise, such as is shown in Fig. 234, and to twist the end firmly with pliers. The Western Union joint is shown in the lower portion of Fig. 247, and the method of making the jointure in Fig. 240. The two wires are gripped by come-alongs, and drawn up with block and falls, as shown, in order to prevent an insecure joint. The two ends are then turned about one another two or three times, and the splice completed as in the former joint, with a pair of pliers. Both these splices are secure and serviceable for telegraph use, but on telephone lines must be soldered or securely wrapped with tin-foil and tape. Soldering, however, is the best practice, being applied, as indicated in Fig. 247. The object of soldering is to overcome the difficulties due to loose contacts at the joints.

Sleeve Joints.—The most approved method for making the joints of telephone lines is by the use of some form of sleeve, such as is shown in Fig. 241. This consists of two copper tubes of the required length, and of sufficient inside diameter, to admit the ends of the wires to be joined, fitting tightly. The tubes are then gripped with a tool, shown in Fig. 246, and twisted around one another, so that the wires are securely joined and locked, as shown in Fig. 241 or in Fig. 247. Another form of joint, also widely used, consists of a sheet of copper bent S-shaped, instead of two tubes. The method of effecting the jointure is similar to that just described. After thus twist-ing on the sleeve joints it is frequently the practice to solder on the wires at either end of the sleeve. Actual tests have demon-strated the fact that the tensile strength of such a joint may be nearly doubled by this method of soldering.

FIG. 245.—Tower wagon for use in repairing pole lines.

CHAPTER TWENTY-FOUR.

WIRE TRANSPOSITIONS ON A POLE LINE.

Transpositions on a Metallic Line.—As has already been stated, the telephonic current is often seriously affected by electrostatic induction from other telephone and telegraph lines, and also from power circuits, owing to the fact that the surfaces of the wires form, as it were, so many charging plates of a true electrical condenser, with the intervening air as the insulating layer or dielectric. The telephonic current changes the potential of its own charging surface as frequently as it alternates, and this fact in itself is amply sufficient to account for a vast weakening of the current before it reaches its destination. The only practicable method for overcoming this annoyance in pole lines is by the arrangement known as "transposition," which is, briefly, the practice of regularly shifting the relative position of the two limbs of each circuit as regards other wires in the same pole system. For short lines and pole systems with only a few wires it is not necessary to transpose very frequently. On longer lines it has been found amply sufficient to transpose once every quarter mile ; that is to say to change the relative position of the wires of the different circuits at posts situated about that distance apart. This does not mean, however, that each pair of wires is transposed so often, but that on ordinary sized systems the transposition of some one circuit is amply sufficient to secure balanced relations and effectually counteract the effects of cross induction. It is a matter which must be carefully calculated and planned in each particular instance in order to secure the best advantages.

Method of Making Transpositions.—The usual practice in America is to use transposition insulators, which are either double insulators, one being screwed to the pin above the other

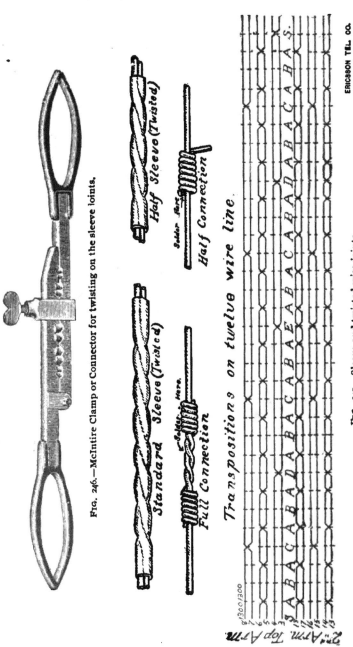

FIG. 246.—McIntire Clamp or Connector for twisting on the sleeve joints.

Half Sleeve (Twisted)

Half Connection

Standard Sleeve (Twisted)

Full Connection

Transpositions on twelve wire line

2nd Arm. Top Arm.

ERICSSON TEL. CO.

FIG. 247.—Sleeve and twisted wire joints.

FIG. 248.—Diagram of transpositions on a 12-wire pole system.

or else such caps as are shown in Fig. 249. Such insulators are intended to act as circuit breakers, the particular wire to be transposed being cut and "dead ended," or tied around, on both the upper and lower grooves of the cap. The free end of each length is then passed back and around the insulator and twisted, or sleeve jointed, to the other limb of its own circuit. With copper wires it is customary to use such sleeve joints as have already been described, but it is necessary only to twist iron wires which are to be soldered as shown in Fig. 247. The plan thus followed may be readily understood from Fig. 251.

FIG. 249.—C. S. Double Transposition Insulator.

Frequency of Transpositions.—The frequency with which transpositions are to be made, and also the relative positions, depend on two considerations: the number of wires strung on each cross arm, and the number of cross arms on each pole. In short it is a consideration of the number of wires strung in the pole system. Figs. 248 and 250 show two different plans of wire transposition, the first for a 12-wire 2-arm system, the second for a 40-wire 4-arm system. In both diagrams the linear spaces between the crosswise dotted lines equal 1,300 feet, or one-quarter mile approximately, and the length of the section shown is about eight miles. A little study of these diagrams will

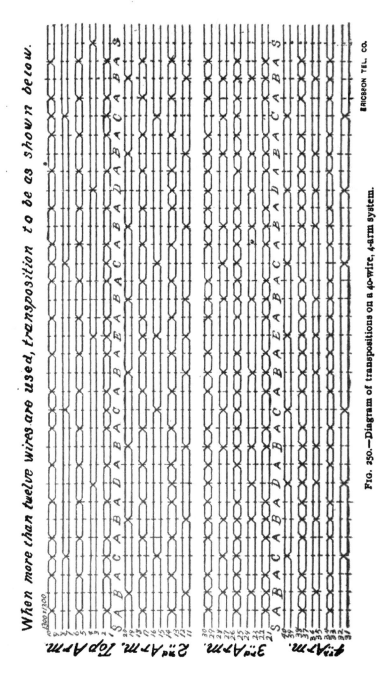

FIG. 250.—Diagram of transpositions on a 40-wire, 4-arm system.

ERICSSON TEL. CO.

show that the transpositions are so made as to insure the fact that no two similarly spaced lines, as, for example, 3–4 and 7–8, are transposed at the same cross arm. Further, the figures show the plan of so transposing all the lines as to end the section, here eight miles long, exactly as it began. In other words, counting back from the point marked *S*, on the right of the figures, we find that the arrangements are in the same order as from the left-hand *S*, and that the same is true of counting in either direction from the points marked *E* at the center of all the diagrams. Thus the same arrangement is repeated within every four miles of the line, the transpositions being made with regular system and the line perfectly balanced, even

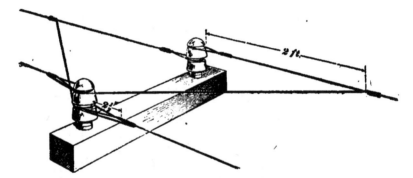

Fig. 251.—The most approved method of making a transposition.

though several hundred miles in length. In some cases, however, as, for example, in the stringing of toll lines, it is essential in order to fully neutralize the numerous disturbing influences that all lines of a pole system begin and end in precisely the same relation; that is to say, that as the first pole is the *S* pole, so also is the last of the series.

General Rules in Transposition.—Although, as has been stated, the plan of transpositions for each separate line must be carefully mapped out beforehand, always taking into consideration the conditions named above, there are some general rules to be observed in spacing the line wires. Thus for a single line pole system it is sufficient to transpose the wires once in every mile. For larger systems new conditions must be considered. Thus a writer in the *Ericsson Series* of telephone pamphlets, referring to the diagrams of transposition, lays down the following principles: "When more than four arms are used, the 5th, 6th, 7th and 8th [arms] should be transposed like the 1st, 2d, 3d and 4th respectively, excepting that at the 1st, 3d, 5th, 7th and other odd *S* poles all pairs except 3 and 4 should be transposed. If only two six-pin cross arms are used, then the wires on the top cross arm should be transposed like 3, 4, 5, 6, 7 and 8, and those on the second should be transposed like wires 13, 14, 15, 16, 17 and 18." The latter correspondence is indicated by the numbering of Fig. 248.

The English Method of Transposition.—In England it has been the practice from the earliest days of telephony to transpose the wires of a pole line on a different plan from that just described. Instead of "dead ending" each wire at stated intervals, and making cross connections to the other limb of the circuit, the wire is transferred from its starting position on one pin and cross arm to another pin and cross arm at the next pole, so that the two wires of a circuit make complete twists around each other through succeeding definite intervals. Thus on a four-span system the four wires make a complete twist once in every four poles. This may be illustrated by a diagram, the letters indicating the several wires and their positions:

First Pole.	Second Pole.	Third Pole.	Fourth Pole.	Fifth Pole.
A B	C A	D C	B D	A B
C D	D B	B A	A C	C D

This diagram illustrates a 2-arm, 4-wire system through its cross-connections on five successive poles. Each cross arm carries two wires, which are regularly cross-connected, as illustrated by the rotation of the letters, A, B, C, D.

The fifth pole is therefore arranged precisely like the first, d the rotation begins again. Such a method is eminently satisfactory for neutralizing the effects of electrostatic induction and other disturbing influences, but it makes a far less sightly line and is more difficult to string and repair. It presents, however, what is after all the true theory of wire transposition and line balancing, for were it practicable to insulate and twist together the two wires of every circuit throughout their entire length there would be no further need of planning for transpositions or cross-connections of any kind.

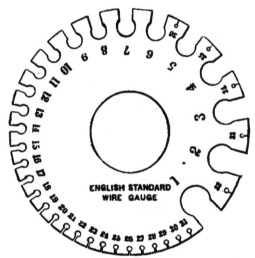

FIG. 251a.—Circular wire gauge, showing approximate dimensions of the wire measures on the Standard, or Birmingham, Wire Gauge System.

CHAPTER TWENTY-FIVE.

TELEPHONE CABLES AND THEIR USE IN UNDERGROUND AND POLE LINES.

Construction of Cables.—The term "cable," as applied to the arrangement of wires fitted for use in an underground or enclosed aerial telephone line is largely a misnomer since, apart from the fact that each such system consists of a number of conducting wires laid or twisted together, there is no resemblance to the structure commonly so called. Telephone cables were devised to meet the conditions incident upon the necessity of running lines underground, particularly in large cities where the law requires it. They might as correctly be termed telephone pipe lines, for such in reality they are. As constructed at the present time, a cable consists of a length of lead pipe through which is drawn a number, generally 100 pairs, of conducting wires. The lead and tin covering serves not only as protection against moisture and other harmful influences, but also as a shield against inductive disturbances, the surface of the pipe absorbing and holding most of the electrical charges likely to interfere with the telephonic current. The electrostatic capacity of the so-called "conference," or standard, cable is usually rated as .8 micro farad.

All the conducting wires are carefully insulated by being wrapped about with prepared paper coverings, and each pair is twisted together as shown in Fig. 252. This "dry" insulation is preferable to some former practices of surrounding the wires with paraffine or other readily fusible material, since the insulation is in no way seriously affected by any degree of heat that the lead sheath can withstand. The wires are thus less likely to be short-circuited with one another, or grounded by electrical contact with the metal of the sheath.

Twisted Pairs of Wires.—As we have seen in the last chapter, the best method for overcoming electrostatic disturbances is to twist the two wires of a circuit around each other through the whole length of the line. This is perfectly practicable in cable construction, and as fully as possible neutralizes any outside electrical disturbances, either between the different circuits or such as leak through the inclosing sheath.

Testing and Connecting a Cable Line.—In connecting a cable line formed of a number of cable lengths, each of which must be unwound from the large reel on which it is shipped, it is particularly essential that no moisture should be allowed to reach the insulating covers of the wires. Such an accident would involve short circuiting or interference, and necessitates the cutting away of such lengths of the cable as are found so affected. The presence of moisture or other causes of

FIG. 252.—A twisted pair of insulated wires, forming two sides of a single circuit, as in telephone cables.

"grounded" or interfering circuits may be discovered by testing each wire in a cable length before it is spliced. This may be done by connecting a battery to one end and an ordinary electric bell or a galvanoscope to the other, when the test is made for "continuity," or by connecting the same end terminal of each wire of a circuit, the one to the battery, the other to the bell, or the wires of each circuit with those of every other, in succession, when the test is made for crosses and grounds. These are only simple tests, various conditions demanding other and more exact methods.

Splicing Cables.—The testing completed, the splicing of the cable lengths proceeds. This involves tight connection of both wires and sheath. Each twisted pair of wires is carefully selected out, untwisted and "skinned" through a short distance; then carefully twisted, each wire to its correspondent in the next cable length, and separately rewound with insulation. The wires are then laid together as compactly as possible, being first carefully " boiled out " by ladling boiling paraffine over the joint until all traces of moisture are eliminated, as indicated by the absence of air bubbles in the paraffine. Next, the joint is carefully wrapped with cotton wadding, which layer is also "boiled out" in the same manner and for the same purpose, and on the completion of this task the lead sheath is joined by an ordinary "wipe joint," such as is made at the joints of lead water pipes in house plumbing. As soon as two lengths have been spliced the line men may proceed to draw the cable through another section of the conduit.

Separating the Circuits.—In order to determine the continuity of any given twisted pair of wires in a cable, either for the purpose of connecting a cable to a pole line or determining exactly which subscriber's circuit it represents, each pair must be tested through on the completion of the line, and the identity of each marked at convenient places, as on cable terminals. In this case the pair is generally designated by its proper number, as, for example, " 120 and mate."

Method Followed in Interior Work.—Where a large number of wires are bunched into cables, as in the construction of intercommunicating systems and in some exchanges, it is customary to make the insulations of the different wires of various durable colors through a series of rotations. Thus the line wire has one designation and the test wire another. The rotation is shown in the following table:

FIRST SERIES.		SECOND SERIES.		THIRD SERIES.	
LINE.	TEST.	LINE.	TEST.	LINE.	TEST.
Red	White	Red	Red and White	Red	Blue and White
Blue	"	Blue	"	Blue	"
Green	"	Green	"	Green	"
Brown	"	Brown	"	Brown	"
Yellow	"	Yellow	"	Yellow	"

By striping the test wires in all possible combinations, then combining the striping of the line wires also, a large number of wires in a system may be perfectly separated into the proper pairs without difficulty and spliced or connected accordingly.

Underground Conduit Construction.—The lead-sheathed cables of telephone lines are not buried in the ground, as are water and gas pipes, but drawn through regularly constructed conduits, composed of a number of pipes laid one upon another. These pipes are made of earthenware, shaped so as to fit closely from end to end, of specially prepared wood—these are called

FIG. 253.—A lineman's testing outfit, including telephonic, signaling and battery apparatus.

"pump logs," from their resemblance to the wood pipes used in old-fashioned pumps—or are built up of cement and concrete, each successive layer being made by cementing around iron gauze semi-elliptical moulds. In making conduit lines with vitrified earthenware pipe lengths, the ends of each pair are carefully cemented and the joint is wiped clean inside, so as to avoid ridges that might injure the lead sheathing, by drawing an instrument called a "mandrel," a length of piping of the same size as the conduit bore, carrying at its further end a flange of rubber of a somewhat larger diameter, which, when drawn through the newly cemented joint, wipes it smooth. The mandrel is placed in the first length of pipe laid down and is drawn forward with a hook as fast as the cement joints are made. At stated intervals, generally at about the length of the standard cable section, the conduit system ends in a manhole, a small cemented room underground, so that the ends of the cables may be brought together and spliced. In laying a cable the first process is to pass a length of heavy flat steel wire, commonly called a "snake," into the designated duct, and push it through to the next manhole. It is there pulled through, carrying a rope attached to its end, by which in turn the cable length is drawn into place.

Overhead Cable Construction.—In some cases, particularly on short lines exposed to inductive disturbances from power and other electrical circuits, it is customary to string the cables on poles such as usually carry the bare conducting wires.

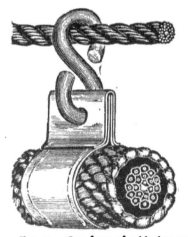

It is not necessary, however, to insulate the cable in any way; consequently it is merely hung to a supporting wire rope or cable, called the "messenger wire," being attached either with some form of hanger, such as is shown in Fig. 254, or by loops of tarred marline. The marline is sometimes wound over the cable and messenger wire from a bobbin such as is shown in Fig. 255, but as frequently it is merely wound on by hand. Cables used in such overhead construction differ in no essential

FIG. 254.—One form of cable hanger for suspending an aerial cable to the messenger wire.

particular from those just described. Both consist of bundles of wires, the pairs twisted together. The size most often used is Number 19, B. & S., which is about .03589 inch in diameter, weighs 20.7 pounds, and has a specific resistance of about 8 ohms to the mile.

Separating and Connecting Cables.—When it becomes necessary to transfer a cable line to an ordinary pole-strung line, or the opposite, or when a cable line is to be connected to the central exchange, some form of the general device known as a "cable terminal" is used. There are many patterns of cable terminal, but all constructions for this purpose are combinations of pairs of binding nuts, most often associated with carbon and fuse protectors, for receiving and connecting the pairs of wires in the cable with the two sides of pole-strung circuits. Fig. 258 shows a typical form of cable terminal of the type intended for an exchange terminal. It consists of a long iron box, which is to be mounted on a slate base, and carries double rows of connecting nuts, combined with carbon and heat coil protectors, down either side. At each end is a short section of brass

FIG. 255.—A form of reel or bobbin used for securing a suspended cable to the messenger wire.

tubing, to which the lead and tin sheath of the cable is soldered. Inside the box the sheath is cut away and the wires are "fanned out," or separated, just as the several sticks of a spreading fan, and attached, each pair to a given pair of nuts. To each pair of nuts are also attached two other terminal wires, which enter the box from above, and are likewise gathered into bundles or cables, which are led to the cross-connecting board, to be attached to the switchboard. This arrangement may be understood from close study of the cut. Fig. 257 shows such a terminal box closed and ready for attachment to its base.

FIG. 256.—The Cook pole-top cable terminal for connecting a cable to an ordinary wire line.

Pole Terminals.—Terminal boxes intended to be attached to poles and to effect connection between an underground cable and an overhead wire line are constructed on the same general principles—the cable being let in and soldered at the base, and the wires attached, bunched together and brought out to the cross arms at the top.   Such a terminal is to be enclosed in a

FIG. 257.—A cable terminal equipped with heat coils.

wooden box on the side of the pole, after being first sealed with a rubber gasket.

Another type of pole terminal is shown in Fig. 256, the Cook pole-top terminal.   As its name indicates, it is intended to be secured to the top of the pole. It consists of a circular cast iron box, as shown, into the bottom of which the cable is led up, being then fanned out and the wires connected to the connecting posts, with the line wires, through suitable air-tight fuse plugs. Each circuit connection is suitably numbered, as shown. After the arrangement is complete a quantity of unslacked lime is placed within the box, and the cover is screwed on with a rubber gasket, as in the other type of terminal. A copper cover is then placed over the terminal box, as protection against the elements.

Exchange Terminals and Distributing Boards.—Fig. 259 shows the type of distributing board furnished with the exchange equipment of the Sterling Electric Co. It consists of a case containing a number of such cable terminals as are shown in Fig. 258, each equipped with suitable protectors. The switchboard wires are bunched together and led out at the top of each terminal box to the proper section of the switchboard, where they are again fanned out and attached to the jacks. In large exchanges it is necessary to employ rather complicated cross-connecting boards, frames supporting shelves for the cables and carrying fused connectors, to which the switchboard wires are attached at one end and the exchange cables at the other. Thus whenever any shifting of the line connections is necessary, the wiring of the switchboard need not be disturbed.

Arrangements for Different Circuits.—In telephony grounded circuits are now seldom used for the speaking current, except on short lines. Consequently, most of the principles of construction hitherto laid down apply to full metallic circuits. In ground return circuits only one wire is strung on poles, the earth forming the return, but it would be impracticable to enclose numbers of such wires in underground cables, since the effects of cross induction, as well as other electrical disturbances peculiar to such circuits, would render talking service nearly, if not entirely, impossible.

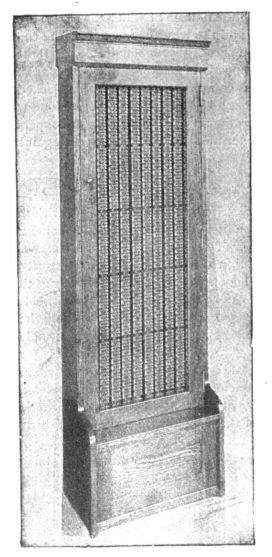
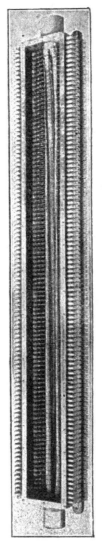

FIG. 258.—Section of a cable terminal box, showing the method of connecting the separate wires to the binding nuts and distributing wires.

FIG. 259.—Type of exchange distributing board. It is a combination of such terminal boxes as is shown in Fig. 258.

Common Return Circuits.—In some cases where it has been found inadvisable to string full metallic circuits for all lines, a method known as the ''common return'' has been adopted. It consists, briefly, in providing the line wire for each station with a common wire for return, instead of the earth.

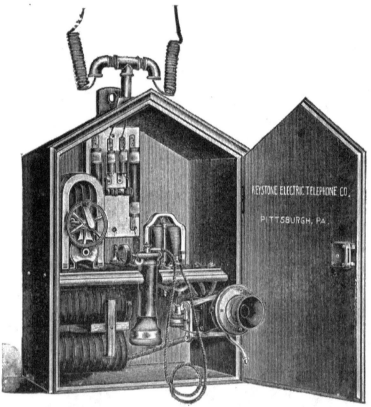

FIG. 259.—A street telephone station, set for the use of the police, fire department or street railway companies. All the component parts of a station apparatus are here shown, although arranged in more compact form than usual.

This kind of circuit works very well with a limited number of stations being handled at a single-wire switchboard, such as has been described in connection with grounded circuits, provided the return wire be of sufficient diameter to insure individual return currents and prevent, as far as possible, leakage in multiple through the wires and apparatus of other subscribers. It is unnecessary to make it of too great diameter, and when, for line wires, Number 12, measuring .0808 inch diameter, is used, a Number 8, of .12849 inch, has been found sufficiently large for common return wires on most circuits.

FIG. 259a.—Circular gauge, showing approximate dimensions of the wire measures on the American, or Brown & Sharpe, Wire Gauge System.

CHAPTER TWENTY-SIX.

CIRCUIT-BALANCING DEVICES.

Repeating Coils and Combined Circuits.—When, as frequently happens, it becomes necessary to join into talking connection a grounded and a metallic circuit, when a short metallic line is to be connected with a long one, or when it is desired to neutralize the troublesome effects of local induction, on some certain section of a line, the readiest and most efficient method is to interpose a repeating coil. This device, as we have already seen, is constructed on the principle of the ordinary induction coil in all respects except that the core is so arranged as to form a complete ·magnetic circuit by being twisted about the windings so as to completely envelope them. Its use is, briefly, to balance connected lines, wherever it may be placed. In connecting a grounded circuit to a metallic, the method is to attach the line wire to one end of one winding of the coil, the other end being grounded, and to attach the terminals of the second winding each to one of the limbs of the metallic line. In the same manner, by the use of two coils, a grounded line may be transformed into a metallic through a noisy section of its length, all the troublesome effects of induction from telegraph and power wires being neutralized. The same is true when a long metallic wire is connected to a short one. Without a repeating coil the combination would be almost intolerably noisy. By interposing the coil, however, each of its windings being attached to the two terminals of one of the lines, all foreign noises are perfectly weeded out.

Construction of Repeating Coils.—Repeating coils, like all other telephone appliances, are made in different dimensions to suit differing requirements. For ordinary line connections, however, the following are excellent figures: The core is formed of annealed iron wires, Number 24, B. & S., and is made of sufficient length to completely envelope the two windings, being bent around them on every side. Both windings are composed of Number 31, B. & S. silk-covered copper wire, of .008928 inch diameter, the primary being wound to a resistance of about 200 ohms, thus making the length about 1,500 feet. After winding the coils and the core the instrument is either clamped to its

FIGS. 260-261.—Two forms of telephone repeating coil.

FIG. 262.—Ericsson five-terminal translator for use in multiplex telephone circuit arrangements.

base with metal straps, as shown in Fig. 260, or inclosed in a suitable sheath, as in Fig. 261. Both ends of the two windings are brought out at the binding screws.

Repeating Coils on Special Circuits.—Although for ordinary exchange work, such as the connection of a grounded and a metallic, or a short and a long circuit, repeating coils are sometimes made with both windings of equal dimensions, as regards length and size of wire and resistance, the requirements in other cases demand such a degree of inductive transformation as can be obtained only by making the windings of differing dimensions. Figs. 263 and 264 show two instances in which this latter practice has been found the most serviceable. They illustrate specifications for line work furnished with the Ericsson

Swedish translators, and are as follows: The diagram in Fig. 263 shows the method of making a full metallic out of a grounded trunking line through a section of its length by interposing a repeating coil at either end of the metallic circuit. For this purpose Ericsson translators are made with a primary winding of about 180 ohms resistance and a secondary of about 320 ohms, an arrangement which involves the raising of the potential as the telephonic line enters the transformer circuit, and the lowering of the potential as it re-enters the grounded circuit at the

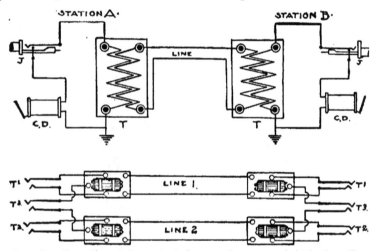

FIG. 263.—Illustration of the method of connecting a grounded telephone line to an intermediate metallic section.

FIG. 264.—Diagram of a triplex telephone circuit arrangement, formed by the use of five-terminal translators.

opposite terminals. Fig. 264 shows an interesting triplex arrangement, as readily applicable to exchange station and private line circuits as to the trunking lines shown in the diagram. By the use of four of the five-terminal translators, such as is shown in Fig. 262, two metallic circuits may be arranged so as to permit of the three telephonic messages being transmitted at the same time, without interference. The arrangement is, briefly, as follows: Each of these five-terminal transformers consists of a primary winding of 170 ohms resistance and of a secondary of 340 ohms. Conse-

quently the circuits designated respectively as T^1 and T^3 have
a total resistance of 680 ohms for the translator secondary
windings in addition to the general ohmage of the line wires.
The line, T^2, which is shown connected to the fifth terminal of
every coil, should have a similar degree of resistance if the
triplex arrangemement is to be maintained without inductive
and other disturbances.   This result is accomplished by attaching the fifth terminal of each coil to the center point of the
secondary winding, so that, entering at this center point of the
two coils at the transmitting end of the line, the current
traverses one-half of the secondary winding of each, going
thence over the test wire of Line 1 and the line wire of Line 2
to the opposite station, where it again traverses one-half of the
secondary windings of both coils to the terminals in the switchboard jack, as shown, or in the subscriber's apparatus.   The
resistance of each secondary winding being, as we have learned,
340 ohms, one-half of each would give 170 ohms, with a total
of 340 ohms at each end of line T^3, or a total of 680 ohms for
the entire circuit, in addition to the general line resistance.   By
this arrangement the lines are kept perfectly balanced, and
there is no interference, cross talk or leakage possible even if all
three are in constant use.

Pupin's Long-Distance Telephone Cable.—Under the
general head of circuit-balancing coils it is proper to place
Prof. Michael I. Pupin's recent invention, which, as is believed,
has solved the problem of long-distance telephony beyond 1,000
or 1,500 miles, hitherto considered the limit, and made possible
transatlantic telephone communication.   Hitherto the greatest
obstacle in the way of achieving the latter feat has been the
high static capacity of the cable.   Consequently, in view of the
fact that better transmitting instruments have seemed impossible, various inventors have busied themselves with the problem
of devising improved cables, in which direction all have been
unsuccessful.   It has been demonstrated that the E M F due to
the self-induction of a circuit lags 90 degrees behind the active

E M F, while that due to the capacity is 90 degrees in advance, and experiment has demonstrated that the one may be made to neutralize the other. The situation as applied to long-distance telephony has been well expressed by Kempster Miller as follows:

"Unfortunately for long-distance telephony such a balancing of self-induction against capacity can be obtained only for one frequency at a time. To thus tune a circuit for one particular frequency would render that circuit capable of transmitting efficiently one particular frequency of vibration, while the requirements of telephony are that all frequencies within the range of the human voice shall be transmitted with equal facility. Again, and unfortunately, it has been found impossible to neutralize distributed capacity with anything but distributed self-induction, and this has not yet been accomplished in practice."

Choking Coils in the Circuit.—After an elaborate series of experiments Prof. Pupin has demonstrated that the static capacity of a conductor may be offset by its conductance by the use of choking coils distributed at intervals of every eighth of a mile. By the use of these coils the circuit may be completely "tuned," and all "blurring" of the speaking current effectually neutralized. His coils are of the ordinary pattern used with electrical circuits, consisting briefly of a coiled wire of high self-induction, containing a laminated core. Both wires of a circuit are passed through such coils, the best arrangement being, probably, to superpose the windings, as in ordinary transformers. In this respect a choking coil differs in nothing from the ordinary transformer or converter, any form of such instrument giving the desired effect when its secondary winding is left on "open circuit." Prof. Pupin's invention permits the use of steel wire in the cable instead of copper, allowing also a much smaller size of conducting wire, and in these advances on current practice cannot fail to revolutionize telephonic construction on long-distance circuits.

CHAPTER TWENTY–SEVEN.

METHODS OF MEASURING WIRE.

Wire Gauges.—There are two leading gauges for giving standard sizes to the diameters of wire—the Birmingham Gauge (B. W. G.), still used in England for all kinds of wire and in this country principally for iron wire, and the American Wire Gauge or Brown & Sharpe (A. W. G. or B. & S.), used for copper wire. The size of any given specimen of wire may be determined either with a circular gauge, such as is shown on

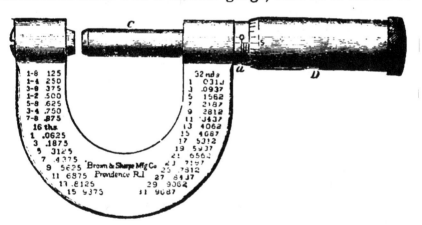

FIG. 265.—Micrometer screw for determining the gauge size of wires.

pages 303 and 314, or by a micrometer screw gauge, shown in Fig. 265. In the former case the wire ends are passed into the holes until the nearest fit is found. In the latter the screw, having forty threads to the inch, is turned until it touches lightly the end of wire placed against the anvil. The collar on which the screw turns is graduated into tenths of an inch, each subivided into four parts, so that, knowing the number of turns

BROWN & SHARPE WIRE GAUGE.

Data on the diameters, weight and ohmage of copper wire.

Gauge Number	[SIZE. Diameter in Mils.	Square of Diameter or circular Mils.	WEIGHT AND LENGTH. Grains per Foot	Po'nds per 1000 Feet.	Feet per Pound	RESISTANCE. Ohms per 1000 Feet.	Feet per Ohm	Ohms per Pound.	Carrying Capacity, 2,000 Amperes p.sq in. section. Amperes
0000	460 000	211600.0	4477.2	639 60	1.564	.051	19929.7	.0000785	430
000	409 640	167804.9	3550.5	507 22	1 971	.063	15804.9	.000125	262
00	364 800	133079.0	2815.8	402.25	2.486	.080	12534.2	.000198	208
0	824 950	105592.5	2236.2	319.17	3.133	.101	9945 3	.000815	165
1	259 300	63694.49	1770.9	252 98	3.952	.127	7882.8	.000501	130
2	257.630	66378.22	1404.4	200.63	4.994	.160	6251 4	.000799	103
8	229.420	52633 53	1113.6	159.09	6.285	.202	4957.3	.001268	81
4	204 310	41742.57	883.2	126.17	7.925	.254	3931.6	002016	65
5	181.940	33102.16	700.4	100.05	9.995	.321	3117.8	.003206	52
6	162.020	26250 48	555 4	79.34	12 604	.404	2472.4	.005098	41
7	144.280	20816 72	440.4	62.92	15.893	.509	1960.6	008106	32
8	128.490	16509 68	349 8	49.90	20.040	.643	1555.0	.01259	26
9	114.430	13094.22	277 1	39 58	25.265	.811	1238.3	02048	20
10	101 890	10381.57	219.7	31.33	31.567	1.023	977.8	.03259	16
11	90.742	8234.11	174.2	24.59	40.176	1.289	775.5	.05181	13
12	80 808	6529.93	138 2	19.74	50.659	1.626	615.02	.08237	10 2
18	71 961	5178.89	109 6	15.65	63.898	2 048	488.25	.13087	8.1
14	64.084	4106.75	86.67	12.41	80.580	2 585	886.80	.20830	6 4
15	57 068	8256 76	68 83	9 84	101 626	3 177	306.74	.33188	5 11
16	50.820	2582.67	54.67	7.81	128.041	4 552	243.25	.52638	4 01
17	45.257	2048.19	43.33	6.19	161.551	5.183	192.91	.83744	8 2
18	40 803	1624 33	34 37	4.91	203.666	6 536	152 99	1.3312	2 5
19	85.390	1252 45	26 50	3.786	264 136	8.477	117.96	2 2392	1 96
20	81.961	1021 51	21 60	3.086	324 045	10 394	96.21	3.3438	1.60
21	28.462	810 09	17.14	2.443	405.497	13 106	76.30	5.3339	1 28
22	25 347	642 47	13.59	1 942	514.983	16 525	60 51	8.5099	1 08
23	22.571	509 45	10 77	1 539	649.773	20 842	47.95	13 334	.80
24	20 100	404 01	8.55	1.221	819.001	26.284	38.05	21 524	.63
25	17 900	320 41	6 77	.967	1034.126	33.135	30.18	84.293	.50
26	15 940	254 08	5 38	.768	1392.083	41.759	23 93	54 410	.40
27	14.195	201.49	4 26	.605	1644.737	52.687	18 95	86 657	.31
28	12 641	159.79	3 39	.484	2066 116	66 445	15 05	137 283	.25
29	11 257	126.72	2 69	.384	2604 167	83.752	11.94	218 104	.20
30	10 025	100 50	2 11	.302	3311.255	105.641	9 466	349.805	.16
81	8 925	79.71	1 67	.239	4184 100	133.191	7.505	557.286	.13
82	7 950	63 20	1 33	.190	5263 155	165.011	5.952	884.267	.098
83	7.080	50.13	1 06	.151	6622.517	211.520	4.721	1402 78	.078
84	6.304	39.74	.847	.121	8264.463	267.165	8 743	2207.98	.062
85	5.614	31 52	.658	.094	10638.30	336.81	2.969	3583.12	.049
36	5.000	25 00	.525	.075	13333.33	424.65	2 355	5661.71	.039
87	4 453	19 83	.420	.060	16666.66	535.33	1.868	8922.20	.031
88	3 965	15 72	.315	.045	22222.22	675.22	1 481	15000.5	.025
89	3 531	12 47	.266	.035	28315.79	851 789	1 174	22415.5	.020
40	3.144	9.88	.210	.030	33333 33	1074 11	.931	35803.8	.015

of the screw, the size of the wire in fractions of an inch may be readily found. The yoke on which the screw turns contains the various B. & S. wire sizes and their equivalents in thousandths of an inch. Thus the gauges may be found at once.

The advantage of the Brown & Sharpe Gauge over all others is that the diameters of wire decrease in the geometrical ratio of 1.26, or the cube root of 2, as the gauge numbers increase. Thus the sectional area of the wire is doubled every three numbers. No other gauge has such a system.

Measuring Wire by Weight.—Of late years the custom is becoming increasingly common of measuring wire by its weight per 1,000 feet, or per mile, instead of by its diameter. This system has its advantages and may be as accurate as the other, since, for a given quality of wire, the weight and diameter must have a constant ratio. Therefore, in buying a wire that weighs so much per mile, we are sure of the desired diameter and cross-sectional area. On this plan wire may be also measured in terms of its resistance or by the "ohm-mile." Two rules, given by Herbert S. Webb, are useful in this particular : (a) "The weight in pounds per mile can be found by dividing the square of the diameter in mils by the constant 62.57 ; and, conversely, the diameter may be found from the weight by multiplying the weight by 62.57, and extracting the square root of the number obtained. (b) The resistance in ohms per mile is found by dividing the constant 890 by the weight per mile." The accompanying table gives the necessary figures for the gauge sizes, weights and ohmage of copper wire.

Fig. 266.—A Dynamo for generating direct currents.

CHAPTER TWENTY-EIGHT.

USEFUL DEFINITIONS AND HINTS ON TELEPHONE MANAGEMENT.

A. W. G.—American Wire Gauge, or Brown & Sharpe Gauge, the standard gauge for copper wire in the United States. The several sizes of wire on this system increase from the smallest in a geometrical progression, whose ratio is 1.26, or the cube root of 2. Thus every third size doubles the diameter 2.

B. & S.—Brown & Sharpe Wire Gauge. American Wire Gauge (A. W. G.) introduced by and named for the firm of Brown & Sharpe, of Rhode Island.

Bridge.—As this word is used in connection with telephonic circuits it refers to the method of connecting any electrical instrument, such as telephonic station apparatus in a party line, between the two sides of the circuit.

Bus Wires: Bus Bars.—The word "bus" is undoubtedly derived from "omnibus," which means "for all things." It is used to designate the wire, rod or bar attached to the dynamo or battery for the purpose of distributing power, which is taken off by bridged connections at every desired point.

B. W. G.—Birmingham Wire Gauge, or standard gauge, the gauge for measuring wires used in England. In America it is applied mostly to iron wire, copper being measured by the American Wire Gauge, or Brown & Sharpe.

Capacity.—This word is used to designate the measure of an electric charge a plate or conducting surface is capable of receiving and retaining, as manifested by its ability to retain a certain degree of electrostatic potential. It is a consideration of importance in connection with line construction, which involves electrostatic conditions similar to a condenser.

Coefficient.—This word means literally, "jointly efficient " or "acting or operating along with." Both in mathematics and physics it involves the idea of multiplication, being such a number as, when multiplied by another expressing degree, will give the difference in some physical condition. Thus, if we know the number expressing the expansion of iron through one degree of heat and multiply by 100 we have its expansion for 100 degrees. The number for one degree is then the coefficient.

Conductance.—The conducting power of a given mass of conducting material of given shape and length. It varies with the cross-shape, directly as the cross-section, and inversely as the length.

Conduction.—The process or act of conducting an electrical current.

Conductivity —The relative power of conducting the electrical current, or providing a path for it. The opposite of resistance.

Creosoting.—This is a term used for a process of preparing wood so as to increase its durability. The wood is placed in an iron chamber from which the air is exhausted, drawing the sap from all the pores. It is then subjected to treatment by high pressure steam, after which crude petroleum is forced into the wood under a pressure of 300 pounds to the square inch.

Dielectric.—A term used for non-conducting substances in general, but more usually for layers of such material placed between the conducting plates of a condenser, like the glass of a Leyden jar, or the paraffine used in ordinary condensers. It permits induction, but bars a current.

Ear.—This word is used to designate any such projecting piece on a mechanical construction as will serve to support, hang or attach some other piece of the contrivance. It often means the same as "lug."

EMF.—Electromotive force. This is a name given to the form of energy which is given off from a battery, and emerges on the circuit in the form of pressure, causing all electrical effects.

Impedance.—This word means literally any form of hindrance or obstacle to the electrical current, true as well as false resistance, but it is generally used as a synonym for the retarding effect of induced magnetism; retardation.

Limb.—This term is used in electrical parlance to indicate either side of a circuit. The line or the return wire is a "limb." The two together are spoken of as the "limbs" of the circuit.

Lug.—In mechanical construction this word is used for any earlike projections, such as are used for attaching one piece to another, either by screws, pins or hooks, or by simple support.

Maximum.—This is the Latin term for "greatest," and is used in scientific language to indicate the point at which any force or fact reaches its highest development or intensity.

Mil.—From Latin *mille*, a thousand. It indicates the thousandth part of an inch and is used to accurately indicate the diameter of wires, etc., in decimal figures. It is .000083 foot. The circular mil is the unit of area and is .78540 sq. mil, or .00000056 sq. in.

Molecule.—This word is used to indicate the hypothetical minute particles supposed to constitute all matter. Chemically a molecule is the combination of several atoms, or ultimate particles, of different substances.

Multiple.—The word multiple refers primarily to anything composed of a number of like parts, or to a multiplication of a thing. In electrical parlance it is used to describe the method of connecting an electrical device between the two limbs of a line, on a "bridge." This is also called "parallel" connection. In the multiple switchboard a number of jacks are thus connected between the two limbs of each separate telephonic circuit.

Normal.—This word is used in electrical science and mechanics to indicate the *resting* condition of a machine. Its usual meaning is correct, true, proper, healthy.

Phase.—This word expresses the form of a wave in oscillatory motion at any given period of time, or the comparison of that position with the standard position. The complete angle of a phase is 360 degrees.

Positive.—This name is given to one pole of a battery and also to the electrical energy found on it because it seems to be the active agent emerging from the cell, while the " negative " appears to be the return side of the circuit.

Potential.—The power of containing or giving off electrical energy. The current is a phenomenon of the passage of energy from a point of positive to a point of negative potential; the former representing the power to give, the latter the reverse.

Relay.—An electro-magnet, which, when energized, attracts its armature with the result of closing an auxiliary circuit or making some other mechanical effect.

Self-Induction.—The production of an induced current in a circuit by some variation of the degree of the electrical energy, due to the expenditure of energy in creating an electric field.

Series.—The method of attaching cells or electrical contrivances in circuit by passing the energizing current completely through each of them on its way from one terminal of the circuit to the other.

Short Circuit.—A term used to express the fact that a circuit is made between the poles of a battery *short* of the contrivance intended to be reached. Any conductor so interposed 'ween the limbs of a ciruit may produce the result.

Shunt.—This word is akin to "shun," and means literally " a turning-aside from." In electrical parlance it refers to a subsidiary circuit connection by which a part or the whole of the current may be turned from the main line. In railroading it refers to a side track.

Step by Step.—This is an electrical expression to describe a machine or effect operated on the principle of energizing a relay by successive impulses so that a ratchet or escapement may be operated.

Teleseme.—A step-by-step contrivance consisting of two dials, on one of which a hand is turned to any desired point, indicating some particular signal, operating electrical mechanism to bring the hand on the other to the same point.

Thermopile.—An electrical contrivance consisting of alternate layers of dissimilar substances in which a current is generated by applying heat.

Transformer.—An induction coil for reducing the initial electromotive force, or pressure, of an electrical current. It is of use in many branches of electrical science where a variation in potential is necessary.

Vulcanizing.—A process of treating wood by heating in a closed vessel to about 500 degrees, Fahrenheit, with the result of coagulating the sap and rendering the wood more durable.

Fig. 267.—Telephone Extension Bell.

TELEPHONE "DON'TS."

Don't tap on the diaphragm of the transmitter or receiver with a pencil or other article. You may injure the apparatus; the ear piece of the receiver may be removed and an examination of the diaphragm made. If bent, replace it with a new one and screw on the cap until it sets firmly in place.

Don't drop the receiver or throw it down; you are apt to break it if you do. The shell is made of hard rubber and is brittle.

Don't experiment with the interior mechanism if you are not posted on telephony.

Don't talk in a loud voice because you do not hear the speaker at the other end of the line very well. The difficulty may be in your receiver.

Don't expect satisfactory results when your receiver cord is broken, binding post screws loose, or where interior contacts have grown poor from want of attention.

Don't expect your telephone to give satisfaction if the batteries are exhausted or connections at binding posts corroded.

Don't expect your telephone to operate if you have forgotten to hang up the receiver and left the battery on a short circuit for several hours; that is, not until it has recuperated, or you have replaced it with another.

Don't place on top of the machine articles of metal. If you do, your telephone may short circuit and you cannot call out to line.

Don't oil the hinges of the bell box.

Don't open the door out of curiosity and then forget to lock it again.

Don't short-circuit the instrument by jamming a lead pencil between the lightning arrester points.

Don't stand too far from the transmitter while talking. Remember, a good telephone is so constructed that it will not gather up distant sounds and is adjusted with a view to short-range operation. Talk from two to six inches from the mouthpiece, according to the length of line over which you are talking and the privacy of your conversation.

Don't talk loud—it is unnecessary—but talk clearly and not too fast.

Don't blame the telephone if you do not perfectly understand at all times the party at the other end of the line. Remember, that all voices are not alike; some are particularly well adapted to telephone conversation, while others are very unsatisfactory.

Don't complain to the office that your telephone is out of order until you are sure of it. Trouble at the other end of the line will, in all probability, affect your own instrument.

Don't forget to ring off when through talking.

Don't expect to obtain good results unless you do your share in keeping up the apparatus and line.

Don't expect the best treatment in the world at the hands of exchange operators if you have given them occasion to put your telephone on the list of "chronic kickers."

Don't waste the operator's time in useless talk. Remember, there are other subscribers to the exchange who also expect her prompt response to their calls.

Don't lose your patience; you are simply powerless, and loss of temper only makes a bad matter worse. If the exchange is not treating you properly, report it, and if no relief is afforded, provided you are in the right, order your telephone taken out.

—Standard Tel. & Elec. Co.

TELEPHONE TROUBLES.

Bells will not ring.—Cause: Broken wire in bell box; line or ground wire—short circuit if bridging metallic, grounded if bridged ground.

Receives and Transmits a ring feebly.—Cause: Bad connections in bell box, or poor ground—resistance cross if bridging metallic—resistance ground if bridged grounded line.

Rings other bells strongly, but its own bells are weak.—Cause: Ringing magnet weak, or armature adjustment bad.

Rings other bells feebly, but received ring is strong.—Cause: Generator weak, or armature adjustment bad.

Receives a ring, but will not ring its own bells.—Cause: Wire broken in generator, or armature short-circuited.

Rings but cannot talk.—Cause: Broken cord, bad connections or hook does not go up to place—line open if bridged.

Can ring, but can get no response.—Cause: Line badly grounded or broken and grounded; if bridged, line open.

Cannot ring or receive a ring.—Cause: Wire broken in office or line—short circuit if bridged metallic—grounded if bridged to ground.

Two switchboard drops fall or two bells ring together.—Cause: Office wire or line wires crossed.—If common return wire, return wire broken or annunciator ground broken.

Bell rings frequently without apparent cause.—Cause: Swinging cross with telegraph or other lines.

Receiver weak.—Cause: Bad connections, diaphragm bent or dirty, position of diaphragm not correct (should be $\frac{1}{32}''$ from magnet), or permanent magnet weak.

Speech received is strong, but transmitted is weak.—Cause: Speaker stands too far from telephone, or battery is weak.

Speech indistinct with a bubbling, buzzing sound.—Cause: Loose connection at microphone.

Spluttering or grating noise in telephone receiver.—
Cause: Loose connection at battery, transmitter or hook.

Can hear but cannot talk.—Cause: Primary circuit open.

Be sure the batteries are properly connected.—Keep
batteries in good condition, as per instructions posted on same.
Connect one battery wire with the zinc pole of the battery, the
other with the carbon pole of the battery. In connecting two
batteries together, connect from carbon to zinc. Never connect
zinc to zinc or carbon to carbon. Battery zinc must be kept
clean and free from crystals and renewed if badly eaten.

The battery cell should be free from crystallized sal-ammo-
niac. The solution should reach the neck of the jar.

Battery wire connections must be carefully guarded against
corrosion.

Observe the following.—Observe the interior wiring and
repair all slipshod work, such as sagging wires, two wires under
the same staple, etc. Keep the receiver well pressed to the ear
when you are listening. If a line is open after a thunderstorm,
a wire is probably burned off in an annunciator or in the ringing
magnets of a magneto bell on the line.

If the diaphragm is bent or rusty, substitute a new one.
Contact points should be soldered, and all working parts of the
instrument free from dust and corrosion. Cords should be tested
carefully for breaks.

Bells should have wire connections well fastened and points
of contact soldered.

See that all binding post connections are tight.

Examine carefully the points or teeth of lightning arresters
after each storm, to see that plates are not in contact with each
other by fused points. The points should always be separated
about the thickness of a sheet of paper.

Dry ice is an excellent insulator, and a good ground wire in
frozen earth is absolutely worthless.

When bell rings, or persons are heard talking on two or
more lines simultaneously, the lines are in contact, or crossed,
or return ground wire broken. —*Western Tel. Cons. Co.*

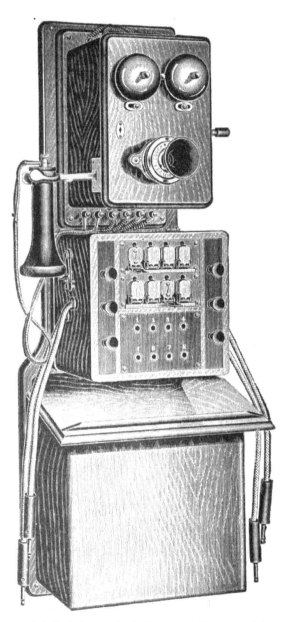

FIG. 268.—Couch & Seeley's combined telephone station apparatus and **switch-**
board for small systems and hotels.

INDEX.